2

AF758318

Disaster Prevention Policies

Series Editor
André Mariotti

Disaster Prevention Policies

A Challenging and Critical Outlook

Patrick Pigeon
Julien Rebotier

ISTE PRESS

ELSEVIER

First published 2016 in Great Britain and the United States by ISTE Press Ltd and Elsevier Ltd

Apart from any fair dealing for the purposes of research or private study, or criticism or review, as permitted under the Copyright, Designs and Patents Act 1988, this publication may only be reproduced, stored or transmitted, in any form or by any means, with the prior permission in writing of the publishers, or in the case of reprographic reproduction in accordance with the terms and licenses issued by the CLA. Enquiries concerning reproduction outside these terms should be sent to the publishers at the undermentioned address:

ISTE Press Ltd
27-37 St George's Road
London SW19 4EU
UK

www.iste.co.uk

Elsevier Ltd
The Boulevard, Langford Lane
Kidlington, Oxford, OX5 1GB
UK

www.elsevier.com

Notices
Knowledge and best practice in this field are constantly changing. As new research and experience broaden our understanding, changes in research methods, professional practices, or medical treatment may become necessary.

Practitioners and researchers must always rely on their own experience and knowledge in evaluating and using any information, methods, compounds, or experiments described herein. In using such information or methods they should be mindful of their own safety and the safety of others, including parties for whom they have a professional responsibility.

To the fullest extent of the law, neither the Publisher nor the authors, contributors, or editors, assume any liability for any injury and/or damage to persons or property as a matter of products liability, negligence or otherwise, or from any use or operation of any methods, products, instructions, or ideas contained in the material herein.

For information on all our publications visit our website at http://store.elsevier.com/

© ISTE Press Ltd 2016
The rights of Patrick Pigeon and Julien Rebotier to be identified as the authors of this work have been asserted by them in accordance with the Copyright, Designs and Patents Act 1988.

British Library Cataloguing-in-Publication Data
A CIP record for this book is available from the British Library
Library of Congress Cataloging in Publication Data
A catalog record for this book is available from the Library of Congress
ISBN 978-1-78548-196-3

Contents

Foreword

The “Earth systems – Environment” editorial field has been enriched by this work by Patrick Pigeon and Julien Rebotier “Disaster prevention policies: a challenging and critical outlook”, a work which deals with issues essential to the evolution of our societies, knowledge and public policy. It frankly addresses problems, often avoided in research, which concretely concern disaster prevention:

– Why are the measures of economic and human losses, as well as existing databases of disasters, so imperfect?

– Why are reports on disaster prevention policies presented with more of a focus on their limits rather than their contributions? Note, for example, that there are not any databases for disasters that have been avoided.

– Why is research and action always collected and evaluated with a focus on the limits of prevention policies?

Therefore, the limits of prevention policies create a paradox for research and action: we are forced to act (and to conduct research), but, at the same time, we are doomed to imperfection (zero risk does not exist for action or for research). When faced with the tragic outcomes of initiatives that continue to disappoint, can we consider the limits of disaster prevention policies as anything other than a sign of failure? How can we think and act in this context of imperfection which can be reduced but not eliminated?

This paradox of research and action on disaster prevention is at the heart of the problem this volume addresses, and the theme falls within the scope of some essential environmental issues:

– It contributes to disaster prevention policies, human development and questions of sustainable development.

– It poses the question of how to integrate scientific uncertainty into discussions and displays of prevention policies.

– It poses the question of the slowness of the evolution of conceptual models despite the fact that researchers are aware of their imperfections.

– It emphasizes the absolute necessity of interdisciplinarity and the integration of the necessary policies, all while underlining the multiple epistemological and management problems.

Clearly, an analysis like this has great environmental, societal and even economic interest:

– How can we further reduce the economic and human losses of major disasters?

– What is the interest and what are the limits for disaster prevention insurance?

– How can very poor populations be included in prevention policies? These populations absorb the primary impact of disasters, something which is underestimated or even hidden for numerous reasons.

We hope that this work by Patrick Pigeon and Julien Rebotier will provide a significant contribution to the revealing of cognitive, institutional, political or social obstacles that affect the evolution of disaster prevention policies as part of the paradox outlined by the authors: thinking and acting in the face of imperfection, for whom, why and how?

All of these questions and uncertainties relating to the prevention of multiple environmental risks that affect and will affect our societies deserve the full attention of researchers. Patrick Pigeon and Julien Rebotier should be thanked for their contribution.

André MARIOTTI, August 2016
Professor Emeritus at the Pierre and Marie Curie University
Honorary member of the Institut Universitaire de France

Introduction

Today, thinking about disasters and disaster prevention requires acknowledging the limits of the research and policies concerning them. Such a statement, which may seem provocative at a time when there have never been so many policies and so much research aimed at preventing disasters, is nevertheless widespread. This is widely recognized by international institutions responsible for preventing disasters, primarily, by the UNISDR[1].

In March 2015, this UN agency organized a conference in Sendai, Japan, and presented a report on the policies implemented over the preceding decades. These policies were based on the Hyogo Framework for Action, which was developed following the Kobe earthquake in January of 1995. The most recent United Nations' report on global disaster risk reduction, known as the GAR Report after its English-language acronym, was also prepared for the Sendai Conference [UNI 15a]. The official declaration that closed the Sendai conference very explicitly recognized the limits of policy and research for disaster prevention. The international bibliography, which supports the GAR Report [UNI 15b], leads to similar observations. These different elements are often the points of departure for international research programs on disasters. All of these reports and programs dedicated to prevention draw their inspiration from several articles considered essential, including Gilbert White *et al.* [WHI 01]. This article posits in particular that the more we know about disaster prevention, the more we lose.

1 The United Nations Office for Disaster Risk Reduction, which has the official name of the International Strategy for Disaster Risk Reduction.

Nevertheless, we can also demonstrate that these policies and research are necessary. This entails looking not just at recorded disasters but at those disasters that have been avoided or even, according to official statements, reduced. What would the reported mortality and economic losses have been if the prevention policies and research on disasters were not there? The response is not easy, but we do not lack for examples proving the capacity of policies and research to manage and understand disasters despite their numerous limits. The still-recent disasters in Haiti and Japan, in 2011, are able to demonstrate just how useful prevention policies are despite their limits. Just comparing the differential in estimated mortality is sufficient. These two disasters demonstrate once again, and in a striking way, that the limits of policies do not only stem from insufficient knowledge of the hazards involved (here, earthquakes), far from it. Knowledge of hazards is certainly necessary, but, on its own, it is not enough to prevent damage. The same applies to knowledge of the elements exposed, such as populations and buildings and their unequal vulnerability.

No matter the type of disaster, multiple experiences demonstrate that knowledge on its own, which we can suppose is better in Japan than in Haiti, is no longer enough to prevent major, deadly or nuclear disasters. The question of disaster prevention implies many choices and their hierarchy. It also addresses the crucial question of the capacity for making a choice, who has that capacity and at what scale. Is it the same in Haiti as it is in Japan? That is why we have centered this book on the political approach to the question.

This book, which is informed and inspired by the participation of its two authors in several national and international research programs on disasters, therefore pursues several goals.

It presents an evaluation of the policies and research concerning disasters and their prevention that is more mixed and balanced than that found not only in the media but also sometimes in the bibliography or in certain presentations from international institutions.

It seeks to shed light on why international and national disaster reports emphasize the limits of policies and research so much while so easily discounting their contributions.

It explains why humanity has no choice but to act and think in the face of imperfection, recognizing the limits inherent in actions and knowledge whatever they may be. It also addresses courses of action for responding to this main objective, which consists of having to think about and manage risks as best we can while also asking ourselves for whom we are doing this.

These different problems overlap on at least one point: although it is necessary to define and quantify disasters in order to understand and manage them, this approach is simplistic. It is true that the approach remains fundamental, even essential. But it is confronted with the limits that every analytical approach encounters. Understanding and managing as best we can requires not considering disasters as such, in isolation, but also as a dynamic component of human societies that have a relationship with their other components. Disaster prevention supposes a comprehensive understanding of human societies in their relationships with their environments. To manage and understand disasters as best we can, we have to be able to picture them not only as such, as disasters can be studied in the absolute. On the contrary, these events are prepared by changes in the relationships that human societies have with their environments. History does not stop after a disaster, contrary to what the traditional representations of having a cyclical basis as the so-called return to normal or the so-called identical rebuilding suggest. Political solutions act simultaneously on societies and their environments, but recognizing their limits leads us to reconsider these policies in turn. In this respect, disasters are fundamentally political. They are in no way socially neutral. Nevertheless, this is what the conventional classification of disasters according to their type, including the category of natural disasters, suggests. This classification is still commonly used today, but we have chosen not to use it here.

The structure of the book shows why it is simultaneously useful and simplistic to understand and manage disasters in isolation and in a segmented way. Fighting against these forms of segmentation, which are found as much in policy as in research on disaster prevention, nevertheless does not make it possible to find a definitive solution. What can be done about these limits of policy and research, which can be adjusted but not completely eliminated? The scale of the ethical and economic issues associated with disaster prevention, especially in a world that is inhabited by more than seven billion people and steeped in socio-economic inequality, demands that these questions be asked.

The first part presents the problematic, and even provocative, paradoxical character of disaster prevention policies today in the form of reports. The often-repeated and accepted paradox, which says that the more we know about disaster prevention, the more we lose, motivates this book. This paradox requires stepping back by going from the disasters themselves to disaster risk. The second part seeks to explain the paradox observed from knowledge and feedback that are mobilized on the disaster risk themselves. They effectively make it possible to identify numerous limits as well as the human capacity to partially reduce them, and, above all, for certain actors, to benefit from it. Part 3 aims to explain why these limits exist, and why we cannot hope to eliminate them while still being able to reduce them. It also proposes courses of action for increasing awareness of uncertainty and imperfection in the political decision-making processes concerning disaster prevention, especially on the local scale. Here, we find the propositions currently being debated concerning the participation of local populations, principally the poorest, in disaster prevention. Not doing so would, according to this understanding, contribute even more to future disasters, in contrast to the principles politically and officially stated.

To meet these challenges, we therefore suggest that the reader starts with the reports on disaster policy that exist today. These official reports can be found in the bibliography or even on the Internet sites of international institutions whether they are more tied to research, such as the well-known Centre for Research on the Epidemiology of Disasters, or to management, such as the Prevention Website[2], which is managed by the UNISDR. At first glance, these reports support an approach to the research on disasters and the prevention policies associated with them that are pessimistic and gloomy.

The second reading, which can sometimes appear in the bibliography, reconsiders these reports. Indeed, we recognize that it is very simplistic to disassociate economic losses from the issues at stake that are associated with them. There are certainly more economic losses, but there is also more to lose. The statistical evaluations available on the mortality associated with disasters also take into account the demographic changes. Admittedly, there are often higher mortality statistics, but, related to demographic changes. It has never been so low, even if the mortality levels still remain too high. We can see the difficulties in establishing such reports, which remain very rough as well as unequally feasible depending on the countries or the units of statistical references. Moreover, evaluation of mortality depends in particular on the period of time taken into account, since a single large disaster can call

2 http://www.preventionweb.net/english/, consulted on January 26, 2016.

into question results until now considered positive. However, despite the methodological limits, such as the limits of statistical data, or even the difficulty finding an agreed-upon definition of what a disaster is, the orders of magnitude available often show a cheering mismatch between potential mortality and the recorded mortality. Although this mismatch cannot be equally verified for everyone, particularly for the poorest, there are also major economic interests associated with disaster prevention. We should remember all the same that corrective measures concerning hazards, such as dikes, dams and spurs, are linked to engineering firms and construction and public works companies. This is the case everywhere around the globe. Insurance and reinsurance companies also benefit from disaster prevention despite the statistical increase in economic losses.

Reports on disaster prevention policies are therefore necessarily very mixed, even if we understand how tricky they are to establish. It is even more legitimate to try to understand why they place more emphasis on the limits rather than on the contributions of prevention policies, which underlie them. It is equally legitimate to try to understand why prevention policies, which nevertheless are multiplying and being refined, are not able to eliminate unpleasant surprises.

In the second part, we propose to expose the explanations for these problems that highlight the number of international and national research programs, as well as international institutions. The main problem of limits comes from deficiencies in knowledge, or even comprehension, of disaster risk. Indeed, contrary to what we might expect to find at first reading, whatever the databases of statistics on disasters, they remain very imperfect. Numerous international networks, such as the IRDR[3] [GAL 15] or even the UNISDR [UNI 15b], indicate it, not to mention the European efforts to compensate such imperfections [MIT 14, DE 14]. If we pursue the question further, we quickly come across the problems that risk definition poses. If we push even further, for example, turning toward conceptual models of the prevention of disaster risk, we then realize that today there is no consensus on any single model. All of them have proven imperfect, even those moving toward a more integrated and systemic reading of disaster risk. Nevertheless, there exists an entire field of research that focuses on reducing disaster risk (Disaster Risk Reduction or DRR). This highly aggravating circumstance is

3 http://www.preventionweb.net/organizations/5686, consulted on January 26, 2016.

mirrored by another very similar research field, which focusses on adapting to climate change (Climate Change Adaptation or CCA).

Under these conditions, we can hardly be surprised by the many limits that the disaster prevention policies themselves encounter. Here again, it is easy to illustrate them whatever the types of disaster taken into account. Moreover, we find them whatever the economic, ecological or cultural differences between the land occupations concerned. The two authors of this book present experiences that underline the similarity between reports, despite the obvious differences between Ecuador and France, or Sri Lanka.

Taking into account the most critical and disparate reports, we can legitimately wonder why it is so difficult to push the limits of management as well as of knowledge, and even utopian to want to eliminate them. We can also ask how we can think about and manage disasters as best we can.

This explains the third part of the book.

Here, we propose to explain again why we find so many major limits to the understanding of disasters and their management. This can contribute to the fundamental problem linked to the structure of the cognitive process, as Bergson and Morin in particular have argued. It is necessary to resort to analysis and division to understand and manage, but this approach is inherently distorted and simplistic. This type of epistemological perspective corresponds with what has been observed, and especially to the difficulties of finding conceptual models of disaster risk, which are simultaneously clear and dynamic while also being integrated. Indeed, as Bergson has also anticipated, and Morin has continued here, the systemic recomposition of the analysis is not enough to eliminate the fundamental problem: the one the analytical breakdown of problems poses. Systemic approaches of the socio-ecological type, such as those supported by the Integrated Disaster Risk Management Society (IDRiM)[4], Renaud *et al.* [REN 13] or researchers from the Resilience Alliance [GUN 02], are therefore not approached from this angle often enough. Yet, Bergson offers only intuition as a solution for avoiding analytical reduction and this is not practical, in any case not in the area of disaster prevention. As for the systemic approach to the question, it poses problems of representation and application that remain significant today.

4 http://idrim.org/, consulted on January 14, 2016.

This epistemological reading does not exclude a more selective evaluation of the limits that disaster prevention policies encounter from numerous institutions or political actors that benefit from risk management and disaster prevention. They are not equal in the face of risk and disaster. Moreover, not all of them participate equally in the management and disaster prevention markets. Some thinkers on capitalism are familiar with the coincidence between practical and opportunistic recovery of risks and disasters by certain actors on the one hand, and the theoretical readings of the failure to uncover definitive solutions to prevention policies on the other hand. It is significant all the same that ecological researchers were able to find relationships with the creative destruction associated with Schumpeter's thinking on the economy. Certain thinkers from the Resilience Alliance school are eager to emphasize these relationships.

In this sense, the critical perspective associated with recognizing the limits that are always present and putting these limits in a social and political context offers the social sciences the opportunity to significantly contribute to disaster prevention. The urgent and very legitimate social demand to reduce disasters and their impact is not foreign to the conditions under which they take shape and that give them their meaning. As this book reiterates, the ongoing search for better prevention is particularly difficult. But in addition, disaster prevention is embedded in social relations, organizing and prioritizing the research, structuring policies and even affecting land occupations. It is therefore also from this perspective of strong social differentiations that we have to look at the approaches of risk and disaster prevention initiatives. Here, we intend to make sense of certain recurring difficulties in disaster prevention, between significant social relations and the many limits of the prevention policies that we cannot completely overcome.

Consequently, we must recognize the necessary and necessarily limited nature of human thought and action in the area of disaster prevention. However, it must also be recognized that these limits are not socially neutral. The question of "how can we act and think as best we can?" is the one we have to, in the absence of any possible alternative solution, ask even more.

Of course, there are already numerous works on this question. Thinking as best we can integrates uncertainty into our thinking about disasters (limits of quantification and limits of political action; [GOD 02]) and complexity [MOR 05]. Here, we also find the contributions of the reading of disasters by socio-ecosystemic thinkers belonging to several groups of researchers. Despite these contributions, the question remains open, as proven by the extreme difficulty of finding a conceptual model of disaster risk that is both complex and representable while also being dynamic and admitting the limits of prediction.

Thinking as best we can also involves putting disaster prevention policies into context, revealing the relationship between the geopolitical forces that drive them. Moreover, here is where the geography of disasters approximates geopolitics, especially urban geopolitics [HUL 06].

Managing as best we can requires being able to represent uncertainty and complexity and to integrate them into the process of making policy decisions on the local scale. If we want to reduce the limits of policy, the type of response prioritised today also requires policies and tools that tend to be inspired by socio-ecological systems. These are multi-actor platforms and knowledge management systems. These solutions are nevertheless strongly disputed by a portion of the research, which the GAR Report [UNI 15b] also reflects. Some researchers feel that knowledge management systems do not integrate local actors into disaster prevention and do not focus enough on more co-constructed and shared organizational knowledge. This point makes it possible for us to better understand why there are so many discrepancies observed between political objectives and their implementation at study sites. It particularly emphasizes the difficult separation of disaster prevention and geopolitical questions. The very sensitive question of knowing how to think about and manage disasters as best we can, because we have to do so, at least on the ethical level, involves power, the sharing of decisions and poverty. Risk and uncertainty have scarcely any meaning for extremely poor populations, which is typical of the reduction of the capacity for choice.

The model that Pahl-Wostl [PAH 09] uses is able to present these political issues clearly. It distinguishes between the incremental evolution of policies without challenging the fundamental structures of the societies that are essentially creating disasters through their own dynamics. Depending on how we look at the model, we can find the mixed results that provide the base of reflection for this book. The results are very pessimistic if we admit that we are continuing to think and manage disasters in a hyper-analytical and segmented manner without fundamentally challenging the engines for creating disasters found within human societies. It is more optimistic if we acknowledge the capacity, even if it is necessarily limited, of policy to prevent a portion of disasters. Multi-actor platforms and knowledge management systems can contribute to it. However, this type of solution, which remains technical, does not fundamentally reevaluate preexisting social and political structures. These solutions can even create new markets related to disaster prevention. On the other hand, the question remains of a knowledge management system's capacity to represent the limits of prevention policies being implemented, and to use these representations as aids for discussion and knowledge sharing during debates about prevention policies. Moreover, how can these tools integrate the decision-making knowledge and above all the needs of actors with reduced capacity, as is the case for the poorest populations?

Fundamentally, none of these solutions will be able to eliminate every disaster, even if they are able to contribute to decreasing the intensity of future damage. Pahl-Wostl's [PAH 09] reflections seem to be consistent with Gunderson and Holling's [GUN 02] reading of resilience. For these authors, resilience does not mean the opposite of disaster: aiming for resilience can, at least in some cases, inadvertently contribute to a disaster, that is not to say that we should not act and think in the hope of reducing disasters as much as possible in the face of imperfection.

PART 1

Disaster Prevention Policies: Paradoxical and Ambiguous Assessments

1

The Rise in Knowledge and Policies Suggest an Increase in Disasters

It may seem entirely paradoxical that we have more and more knowledge about disasters, more prevention policies and more institutions dedicated to the latter, and yet the number of disasters and the associated damage is statistically increasing, at least in certain categories. The opposite can be true in some cases: according to the database of the association International Tanker Owners Pollution Federation (ITOPF, [ITO 15]), the frequency of major disasters associated with accidental maritime pollution and international transportation of hydrocarbons is falling. Despite these counter-examples, which we will return to, it has become more common for the bibliography to envision disaster prevention along with research on solutions to the different types of damage, which are increasing and that could even get worse in the future if they are not fundamentally better understood and managed. This is very clearly the message that is being carried by researchers and the policies concerning adaptation to climate change. For Alcoforado and Andrade [ALC 08, p. 250], "having a better understanding of the role that urban areas play in global warming is urgent for us to combat the negative consequences of global warming and take the necessary measures to reduce its consequences or to adapt to new climactic conditions". According to Downton [DOW 09, p. 13], "we need towns to live, and yet they are killing us". Such a diagnosis requires that we transform towns into "engines of survival" [DOW 09, p. 15] by developing "new types of urban ecosystems and significantly transforming our way of life in a substantial and systemic manner" [DOW 09, p. 14]. The types of adaptation to climate change, and especially their wide acceptance by social actors [SUM 10], are a part of the disaster prevention process. These authors, among many others, presume that if the implementation of policies for adaptation continues to encounter such large obstacles, we must expect the increase in damages associated with hydro-climatic disasters in the decades to come.

Researchers specializing in disaster prevention also emphasize the limits of policy, even in countries such as Japan which are famous for being at the forefront of the field. It is not a coincidence that the framework for action known as the Hyogo Framework for Disaster Prevention explicitly refers to multiple returns on Japanese experiences, especially to the 1995 earthquake (Kobe). Kelman and Glantz [KEL 15a, p. 105] nevertheless do not hesitate to describe the most recent Japanese disaster in this way: "March 11, 2011 showed how much more work has to be done in order to reduce disaster risk, even in the presence of information, financial resources, talent and a 'culture of prevention'".

What foundations do such predictions and reports rest on? First and foremost, they rely on the information that certain databases provide about disasters. This is why we are beginning our presentation of the problem with the latter. As we see a trend in the increase in the frequency of disasters, especially noticeable in the category of what are known as "natural" disasters, we will concentrate the analysis on two databases: EM-DAT and DesInventar. We draw attention to the limits of separating "natural" and technological disasters, which are becoming more and more recognized. The Analysis, Research and Information on Accidents (ARIA) database [BLA 15] "shows accidents and incidents starting in 1992 involving classified installations", but "since 2010, ARIA has integrated other areas: transportation of hazardous materials by road, rail, waterway or channelization, domestic distribution and utilization of gas, mines and quarries, hydraulic structures". Indeed, "the 2014 inventory of technological accidents", has a "flood risk" column. It is also a way of attracting attention to the hybrid nature of disasters, a subject we will be returning to.

1.1. According to the EM-DAT and DesInventar databases, the number of disasters would increase

1.1.1 *What is a disaster?*

Answering this question is necessary in order to quantify disasters and their evolution, so that we can better understand and prevent them. Defining means delimiting, and this approach is essential. It is difficult to analyze, quantify, understand or manage disasters without delimitation. Cambier [CAM 05, pp. 39–40] mentions just how long ago this question was first asked, specifically by Descartes. In the preamble of the *Discourse on Method*, Descartes deplores towns that are "ordinarily so badly plotted", that is, measured with a compass. He regrets how geometrically irregular they are, which impedes their management as well as their control. Cambier comments: "the Cartesian type of town rests on a projection that is supposed to be exhaustive and has a mechanically punctilious layout". Just as with the definition of towns, definitions for disasters exist because it is necessary to define and delimit them in order to manage them. However, these attempts at definition encounter numerous difficulties.

The first difficulty is related to the vocabulary itself, since it distinguishes *a minima* between disaster and catastrophe without always making this distinction precise. As one example, and we have a large choice of them, let us consider Albouy's book, *Le temps des catastrophes* (in English: *The Time of Catastrophes*) [ALB 02]. To establish a report, he refers to the database maintained by the Centre for Research on the Epidemiology of Disasters (CRED). Officially, the database focuses on disasters, but Albouy does not hesitate to translate the English term "disaster", into the French and English term "catastrophe". However, the two terms are not strictly synonymous. The term "disaster" is more neutral in terms of theory than "catastrophe", which was used in mathematics by René Thom. For the latter, "there is catastrophe when there is a phenomenological discontinuity… The word came naturally to me, physicists introduced the terminology of catastrophe, infrared catastrophe and ultraviolet catastrophe for phenomena of a divergence in a series in their calculations. But I wanted this word to indicate that it involved something dynamic, that an underlying dynamic exists" [THO 93, p. 28].

In addition, some researchers are hesitant to use the terms "disaster" and "catastrophe". This is the case, in ecology, for Gunderson and Holling [GUN 02]. They prefer the term "disturbance" or "collapse", also in relation to the work of Diamond [GUN 02, p. 116]. Moreover, the term "disaster" does not appear in the glossary. Catastrophic folds, which refer to thresholds suddenly exceeded in certain ecosystems and which are quantifiable, are mentioned only on pages 207 and 208 of Gunderson and Holling [GUN 02]. Nevertheless, in the theoretical sense, the concept of catastrophe is not used as such by these two researchers. This shows reluctance in the face of variable contents associated with this concept. René Thom himself states that he was criticized by a group of researchers in mathematics for using the concept of catastrophe.

These two examples, among many others, confirm the astonishing lack of consensus on what the precise definition of a disaster or catastrophe might be, despite the evidence of economic and human issues at stake that are universally recognized. Moreover, CRED's database is officially known as EM-DAT (for *EMergency DATabase*), while the institution that maintains and manages it explicitly refers to disasters.

This does not prevent certain authors from seeking to make a distinction. We can mention here the work of Quarantelli [QUA 06]. The latter, who worked at a disaster research center in Columbus, Ohio, argues that disasters and catastrophes are distinguishable according to the dynamics and the social behaviors that are observed during the events. Catastrophe challenges daily life much more, and therefore, also assumes that local populations will more likely need external aid, than disaster: "Even in major disasters, there is no such massive across-the-board disruption of community life even if particular neighborhoods may be devastated as happened in the Mexico City earthquake of 1985, but life in many contiguous areas went on almost normally. Similarly, this was true of the Northridge, Los Angeles earthquake of 1994; for instance, 2,000 people went as usual to the horseracing track in that California area the afternoon of the earthquake". According to Quarantelli, unlike Mexico (in 1985) and Los Angeles (in 1994), where the functioning of businesses was not strongly disrupted, the damage caused by Katrina (in New Orleans in 2005) was a catastrophe. That being said, Mancebo [MAN 06], who nevertheless cites Quarantelli's work, also uses the term "disaster" to describe the same event. Obviously, the distinction that Quarantelli offers, and which attracts Mancebo's attention, is not decisive enough to be strictly reproduced by the latter author.

Without wanting to multiply the attempts at definition, we have recently found another attempt at clarification, which seems to reverse that of Quarantelli, in the introduction of the worldwide report on the evaluation of disasters [UNI 15b, p. 25]. Maskrey and Lavell state that, "in many ways the Indian Ocean disasters were a representation of *dis-astrum* (Latin for 'bad star'), the impact of an infrequent and unexpected natural event of extraordinary magnitude outside of human agency". Therefore, like Quarantelli, they assume that there are several categories of disasters, some of them (disasters in the sense of *dis-astrum*) exceeding the human capacity for management even more, but without making them completely ineffective. In fact, in case of an alert, evacuation of at least part of the population is possible. This was confirmed during the 2011 Japanese disaster in the Sendai region. There is even information on the Internet that takes advantage of the population's feedback in hopes of improving preventive measures. These devices have numerous limits, but they contribute to lowering mortality, even in cases of *dis-astrum* as defined by Maskrey and Lavell.

During a field trip in March 2015 organized by the University of Sendai, one of the authors of this book visited the Yuriage site in the city of Natori, Miyagi. This part of the commune was directly affected by the tsunami, as the photograph taken in March 2015 from the Hiyoriyama Temple (Figure 1.1) shows. According to Suppasri *et al.* [SUP 13], out of 7,000 inhabitants of the Natori commune, 890 dead and 231 disappeared were counted, mainly in Yuriage. The height of the tsunami reached 8 m, and the wave arrived about 45 min after the earthquake. However, mortality would have been even higher without prevention policies and early warning systems.

Figure 1.1. *Disaster according to Maskrey and Lavell [UNI 15a, UNI 15b]. Yuriage, Natori, photograph taken in March 2015 from the Hiyoriyama Temple (photograph by P. Pigeon)*

The tragic trap in which the inhabitants of Yuriage seemed to find themselves also demonstrates the distinction between exposure and vulnerability: a disaster in the sense of *dis-astrum* makes human societies more passive, more powerless and more victimized when faced with tsunamis such as those in the Indian Ocean in 2004, or the coast of Sendai in 2011. In such cases, the socioeconomic differences within societies therefore explain less the differentials of the damage recorded. From this point of view, the 2011 tsunami in Sendai, Yuriage, is more of an equalizer when it comes to mortality, since the population there was more exposed simply by its presence at the affected site. Conversely, with Katrina, "the disaster in New Orleans represented a predictable and tragic *kata-strophe* (Greek for 'downturn'), the tragic finish to a long drama. While Hurricane Katrina was an intense hurricane, it was the historically configured risk in New Orleans, the vulnerability of those left behind and the lack of effective actions to assist them that conditioned the scale of the disaster" [UNI 15b, p. 25]. Here, *kata-strophe* highlights the importance of and above all the deficiencies in human preparation, and the political dimension, that Hurricane Katrina revealed. As a result, an explanation of this type of disaster, *kata-strophe*, brings to light human choices and responsibilities as much in the creation of risk as in the responses provided. Here, exposure is much less significant than vulnerability or socioeconomic differences within the populations involved. With *kata-strophe*, all human beings are not equal in the face of death, especially when it comes to the wealthiest.

We will find that Quarantelli and Maskrey and Lavell's approaches show an underlying internal coherence, but this does not prevent them from differing on the choice of concepts used to express the thought. Maskrey and Lavell's *disaster* is close to Quarantelli's catastrophe, and Quarantelli's *disaster* corresponds more to Maskrey and Lavell's *kata-strophe* (Table 1.1). There is an obvious underlying agreement: an interest in distinguishing between a disaster and a catastrophe depending on the unequal human capacity to prevent and manage events. However, *in fine*, these distinctions are reflected in the choices of concepts that are different and not strictly symmetrical. The polysemy of concepts that reflect the thought is a difficulty. It should be noted that if Maskrey and Lavell consider the Katrina episode as a *kata-strophe*, which makes the socioeconomic inequalities and the differences in the treatment of populations prominent, Quarantelli sees, in his opinion, a catastrophe that completely disturbed social functioning, and that affected everyone as a form of leveling. For these authors, the concepts of disaster and catastrophe do not correspond exactly to a mirror usage where the *kata-strophe* of one (the deficiencies in human preparation and the contrasting consequences among societies and in space) would take on the meaning of disaster from the other (the disturbance of a part of the functioning of societies only).

	Quarantelli [QUA 06]	Maskrey and Lavell [UNI 15b]
Disaster	*Vulnerability is more important than Exposure* Disaster Example: Mexico, 1985. "Particular neighborhoods may be devastated as happened in the Mexico City earthquake of 1985, but life in many contiguous areas went on almost normally" => *Differentiated consequences over the functioning of societies (functional reading).*	*Exposure is more important than Vulnerability* Dis-aster (bad-star) Example: The Indian Ocean tsunami, 2004. "The impact of an infrequent and unexpected natural event of extraordinary magnitude outside of human agency" (p. 25). => *Huge consequences (which affect everyone), considerable reduction of the capacity for choice?*
Catastrophe	*Exposure is more important than Vulnerability* Catastrophe Example: Katrina, 2005. "Even in major disasters, there is no such massive-across the board disruption of community life" => *Huge consequences (which affect everyone), considerable reduction in the capacity for choice?*	*Vulnerability is more important than Exposure* Kata-strophe Example: Katrina, 2005. "The disaster in New Orleans represented a predictable and tragic kata-strophe (Greek for "down-turn"), […] it was the historically configured risk in New Orleans, the vulnerability of those left behind and the lack of effective actions to assist them that conditioned the scale of the disaster (p. 25) => *Social inequalities underlie the differentiated consequences for the society (social and political reading).*

Table 1.1. *Disaster and catastrophe according to Quarantelli [QUA 06] and Maskrey and Lavell [UNI 15b]*

All in all, the literature shows the extreme difficulty of finding a strict separation between the two concepts of disaster and catastrophe, even though it is possible, and even useful, to distinguish between them. Reflecting on the justifications for this distinction reveals some interesting perspectives. For example, a reading for Quarantelli's disaster that is more functional, and another that is more social and political for *kata-strophe* according to Maskrey and Lavell (Table 1.1). In addition, even in cases of disasters in the sense of *dis-aster*, the human capacity for prevention, no matter how reduced, remains and contributes to explaining the differentials in mortality observed; this is what the example of the Sendai region during the tsunami of 2011 shows. As Maskrey and Lavell's interpretation demonstrates, *kata-strophe* shows how much humanity can contribute to the formation of its own disasters.

All of this may rightly seem to be byzantine, and even troubling: the basic analyses are clearly not stable. Frankly, they may even seem contradictory, at least at first glance. There is no consensus on the underlying question of disaster prevention, which is still nevertheless fundamental to humanity.

The differences in definitions between major authors seem to show *in fine* that many researchers and institutions dedicated to disaster prevention have given up trying to specify the difference between catastrophe and disaster. Moreover, large differences according to disaster databases should also be expected.

The latter remark is crucial given that the results of the analysis of these databases preferentially go in the direction of an increase in the frequency of disasters. Such an increase poses, rightly, multiple problems of interpretation and management. It does not seem very consistent with what the distinction between *dis-aster* and *kata-strophe* assumes, which is the human capacity, however limited and unequal, to prevent disasters.

1.1.2. *Even more disasters according to EM-DAT*

Taking into account the difficulties that the literature demonstrates in defining disasters, it is not surprising to find solutions to their statistical categorization that are both heterogeneous and variable according to the databases. The question has already been raised elsewhere, particularly as part of books and reports resulting from programs funded by the European Union: for example, in Menoni and Margottini [MEN 11], Mitchell *et al.* [MIT 14], De Groeve *et al.* [DE 13, DE 14] and Pigeon [PIG 15]. Here, we show an approach to the question focused on the comparison between EM-DAT and DesInventar, the two major databases documenting disasters.

According to CRED [CRE 15], an event can be entered into the EM-DAT database if it meets at least one of the four chosen criteria:

"For a disaster to be entered into the database at least one of the following criteria must be fulfilled:

1) Ten (10) or more people reported killed.

2) One hundred (100) or more people reported affected.

3) Declaration of a state of emergency.

4) Call for international assistance".

These criteria are very close to those that are used in the Canadian Disaster Database, which nevertheless distinguishes in a conventional way between "natural catastrophes" and what it calls "technological incidents" [GOU 15]. Finding very similar definitions involving more heterogeneous databases than EM-DAT draws our attention to the limits of analytical distinction between types of disasters. As we can see, EM-DAT uses two quantitative and two qualitative criteria to define an event such as a disaster and then to integrate it. The first two criteria involve the mortality figures and the people affected. The important information is connected to the crossing of a quantitative threshold, which approaches the intensity of the damage recorded during the event. A disaster corresponds to an event with a stronger comparative intensity of damage. The first two criteria suggest multiple biases in the collection of information, due to the necessity of choosing a threshold. They also note the indirect nature of the collection: the CRED database is in fact a summary of other preexisting databases.

The two qualitative criteria (number 3 and 4) are therefore very useful and complementary. They emphasize on exceeding local management capacity during an event. Surpassing capacity requires turning to external aid, in compliance with the fourth criterion.

We can immediately see that the criteria, even if they are very mixed, show an overall logic. In fact, they develop and specify the basic idea that disaster represents an event that goes beyond the usual conditions for the functioning of a society. Their frequency must therefore remain low over a given period of time. In addition, these criteria also draw attention to the intensity of the damage during the event that is qualified as a disaster, which must be, compared with others, higher. Under these conditions, it is not surprising that the societies involved see their management capacities exceeded. Logically, a spatial reference is added to these events that have a low frequency and a relatively high intensity of damage: the necessity of using external aid to the local society, which the Anglo-Saxon literature readily associates with the "local community". Disaster, by definition, exceeds the management capacity of the local societies simply because the limits of their resources no longer allow them to manage an event that appears catastrophic. Here, we find the etymological sense of a violent overthrow of the normal state, which moreover evokes the definition of *katastrophe* of Maskrey and Lavell [UNI 15b].

A disaster, as defined by EM-DAT, shows astonishing but significant similarities to urban settlements. They also, by definition and in essence, are utopian, if we use Cambier's formulation [CAM 05, p. 51]: "Every town rests on a principle of utopia, in the exact sense of a principle of separation from the constraints of the here and now". No town, in fact, can live purely off of local resources, and this is true even under normal conditions. Disaster also demonstrates, and accentuates, this situation of dependence on external resources, which is structural in urban land occupations. However, this situation of dependence also allows obtaining external aid in case of a disaster, something that clearly underlines the paradoxical character of urbanization.

This suggests that we can neither understand nor prevent a disaster from just the resources of a local society, and therefore, it is also necessary to rely on the outside resources. It also suggests that disaster prevention involves strengthening of local resources in order to attempt to limit the intensity of any future damage. However, prevention at the local scale, as necessary as it is, is not enough to eliminate all damage. This point is even considered to be essential for disaster prevention in the international literature, as we will see later in this chapter. Under these conditions, management of disasters is necessarily more of an urban, and even metropolitan, issue, since it cannot be conducted exclusively at the local scale: it requires analysis and management necessarily at numerous scales. These elements are logically consistent in defining what can be a disaster in substance, and indeed, the principal sources of the EM-DAT database are international. They rest largely on the databases of the major reinsurance companies, and international press agencies. By definition, Munich-Re, Swiss-Re (Zurich), the Reuters Agency (London) or AFP (Paris) are metropolitan. In addition, CRED is located in Louvain-la-Neuve, that is to say Brussels, Belgium.

Summarizing what we have learned about the four criteria that CRED uses to define a disaster, we note that they lend themselves rather well to quantification. Even more so, they make it possible to create distinctions, and even gradients, between detrimental events. Indeed, we see an inverse relationship between the frequency of the events and the intensities of damage associated with them. This statistical property has been observed in several areas, and it indicates the existence of rank-size laws, also called power laws [DAU 13, PIG 10]. The inverse statistical relationship between the frequency of the events and the intensity of the associated damage warrants the question of separating disasters from catastrophes, or, in fact, between types of events that are unequally detrimental. We propose using the principle of the Farmer Curve to represent these distinctions in a simple manner [PIG 10]. They also correspond to the differences that make sense for risk managers, primarily insurers. For the latter, events with a high frequency and a comparatively low intensity of damage do not fall within the scope of insurance. As seen by insurers, these are not in fact officially disasters. They can be managed directly and locally by the citizens themselves, be they insured or not. In addition, their inclusion in the insurance sector would make the activity unprofitable. The principle of the deductible is justified. Events with a lower frequency but a higher intensity of damage fall within the scope of the insurance sector, which, by mutualization, shares the cost of recovery. By contrast, extremely infrequent events whose intensity of damage is very high are beyond the scope of insurance. Here, we find the pertinence of the distinction that Quarantelli makes between disasters and catastrophes: a catastrophe considerably reduces human opportunities for prevention, but this does not mean that human societies remain completely passive. However, their capacity for action and prevention is necessarily much more limited. Catastrophe provides more of an illusion of an exclusively exogenous shock that affects exposed human societies. The limits of prevention also explain why this range of extreme events involves insurance very minimally.

Figure 1.2 establishes these different elements of differentiation, which remain, for now, very analytical and static.

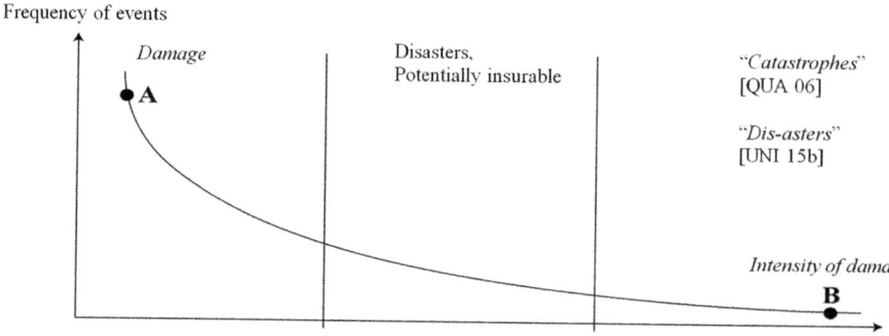

Figure 1.2. *The Farmer Curve and the definition of disasters according to the inverse relationship between the frequency of events and the intensity of the associated damage*

COMMENTS.– *Events with a very high frequency with low-intensity damage (A) do not allow for the insurance principle, which would be unprofitable in these cases. This justifies the existence of franchises. For high-frequency events (A), the insured becomes its own insurer. For low-frequency events with a very high intensity of damage (B), insurance is replaced by national or international authorities. The field of insurance is economically justified by the mutualization of losses associated with intermediate events, which exceed local capacity.*

As we can see, the statistical reviews that CRED and EM-DAT authorize make it possible to show trends over a secular period of time (as shown in

Figure 1.3). At first glance, the report is very clear, so clear that it is integrally repeated in the literature, and it is the paradox at the origin of this book. Indeed, the signature of secular change remains exponential despite the past few years that suggest a reduction in the trend.

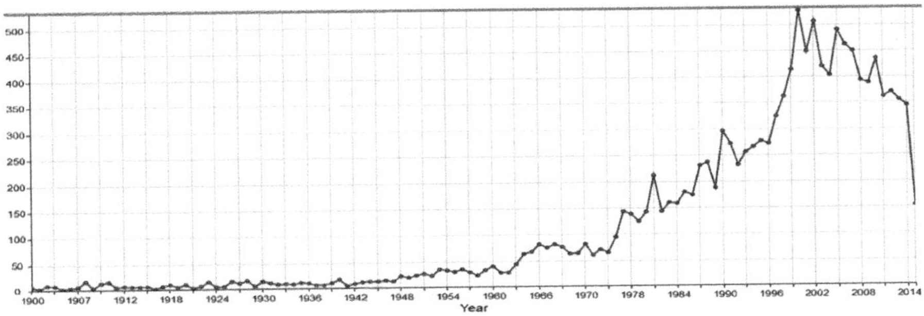

Figure 1.3. *Total number of reported natural disasters between 1900 and 2015. http://www.emdat.be/disaster_trends/index.html*

Nevertheless, the choice of thresholds, as well as the main sources that are connected to the EM-DAT database, draws attention to its limits. We will come back to this. For the moment, we see that the database is clearly geared toward disasters and catastrophes following an approach that is strongly influenced by the economic actors, which are insurance companies and especially reinsurance companies. First of all, its scale is necessarily global, or even national. Consequently, the use of this database by researchers and managers engaged in prevention of disasters poses problems at a more local scale. Nevertheless, this local scale is considered essential in the field of disaster prevention.

It turns out that the change in the scale accentuates once again the pessimistic character of the report. It increases the frequency of disaster that we observe as a result, consistent with what the Farmer Curve shows. However, are those small disasters still disasters? Are we still talking about disasters?

1.1.3. *Even more disasters according to DesInventar, which integrates "extensive risks"*

The limits of EM-DAT justify other databases, such as DesInventar. The DesInventar database was first developed in Latin America with the aim of taking more local disasters into account (La Red, Corporación OSSO and UNISDR, 2002), which some researchers have, unlike EM-DAT, called “small-scale disasters” [LÓP 08]. The database includes events where the intensity of the damage is much lower, and the quantitative criteria seem less strict [DER 09b]. We recall nevertheless that EM-DAT can include an event as a disaster even if no statistical information on mortality or the affected populations is available. The advantage of DesInventar rests in its ability to map the distribution of small-scale disasters in an urban area such as Medellín, in contrast to EM-DAT ([LÓP 11], Figure 1.4). In return, the DesInventar database, as expected in keeping with the Farmer Curve, also shows a much higher frequency of disasters than that seen in the EM-DAT database. It justifies pessimistic readings and even more strongly challenges the reports on policies aimed at preventing disasters.

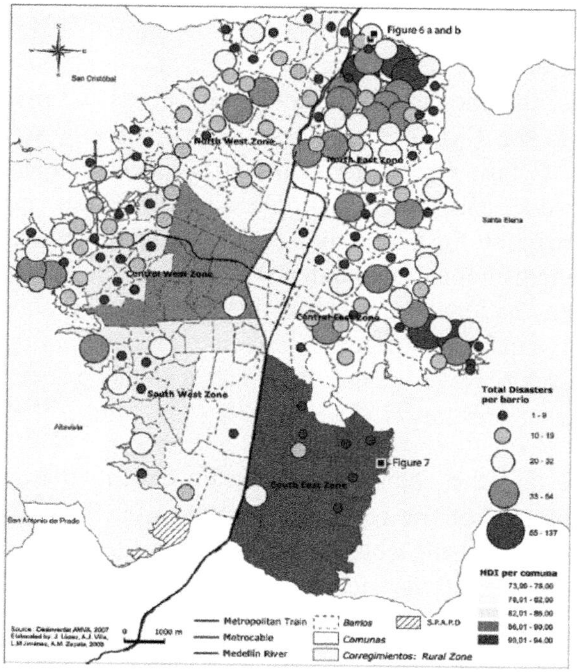

Figure 1.4. *Map of “small-scale disasters” in Medellin, Colombia in the DesInventar database [LÓP 11, p. 576]*

DesInventar's contributions to defining disasters and databases are therefore multiple. They go well beyond the ability to have diagnostics concerning the distribution and evolution of local disasters (this in itself is already a considerable contribution to disaster prevention).

In terms of information, the term "small-scale disasters" or even "local disaster" can appear intrinsically contradictory, even illogical. Indeed, by definition, no disaster can be of small scale (low intensity of damage) or even local. But we see later on that this type of paradox is false. It can be given a logical solution by integrating the differences in scale and the period of time over with which the analysis is performed. Moreover, a solution is suggested in Medellín by another inverse statistical relationship. Indeed, we find an elevated frequency of small-scale disasters in poor neighborhoods often associated with informality, while disasters as defined by CRED first concern consolidated neighborhoods associated with populations that are more socioeconomically prosperous. The frequency of small-scale disasters is as low as the proportion of middle-income households is high. Such a difference cannot be explained only by hydrogeological conditions, which are comparable here [LÓP 08]. It involves acknowledging the capacity of human society to modify its environment, for example, by corrective works such as drains and channelization of watercourses. These corrective works contribute to lowering the frequency of harmful events [PIG 15]. However, reducing the frequency of small-scale disasters does not mean the complete disappearance of any future disasters. This is even less in the case when areas protected by infrastructures are built without taking into account the fact that the disaster risk, even though it has been shifted in time and space, still exists. There are therefore spatio-temporal relationships between DesInventar's small-scale disasters and disasters under CRED's definition. Significantly, managing small-scale disasters, at least in certain cases, inadvertently contributes to disasters as defined by CRED.

Taking small-scale disasters into account, DesInventar reveals the weaknesses of databases, especially where there is only EM-DAT. It reveals how poorly databases like EM-DAT take "extensive risks", as defined by UNISDR, into account.

The weaknesses of existing disaster databases have been identified by UNISDR. Their report on the global evaluation of disasters does not hesitate to devote a whole chapter to it. Indeed, the term "extensive risks" denotes, in

UN terminology, any high-frequency event with an intensity of damage that is comparatively weak, seen from the global scale. As the 2015 report expressly states, "in general, this layer is not captured by global risk modeling, nor are the losses reported internationally" [UNI 15b, p. 90].

Here, we reproduce the figure of the report concerning Indonesia, since it highlights the reduction of the frequency of disasters, which results from insufficiently taking small-scale disasters into account (Figure 1.5).

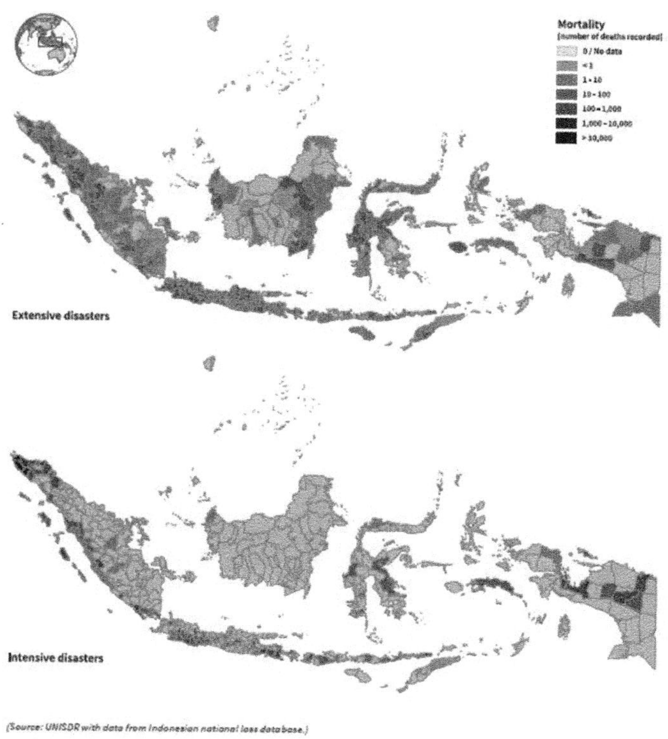

Figure 1.5. *"The different footprints of extensive and intensive disaster loss in Indonesia, 1990–2013". From [UNI 15b, p. 93]. The "extensive risks" underestimated in the statistics on disasters concerning Indonesia*

On the whole, we find that the problems of definitions concerning disasters do not inhibit the existence of databases, which tend to be refined with time, being moreover complementary. However, the report that emerges justifies a very pessimistic interpretation of prevention policies, since it clearly demonstrates an increase in disasters, and even sometimes of mortality and economic losses. It can be enlightening to consult the diagram of the losses involving Colombia, to which Juanita López contributed (Figure 1.6), in UNISDR [UNI 15b, p. 34]. Indeed, it shows explicitly that more detailed information on disasters, for example, that which is available in DesInventar, allows for the adjustment of diagnostics and prevention policies. Nevertheless, it does not imply a limitation of loss, at least at first glance. Such a report is even more troubling in that we simultaneously see a multiplication of disaster prevention policies along with an intensification of research efforts. Figure 1.6 therefore justifies the interest in clarifying this paradox.

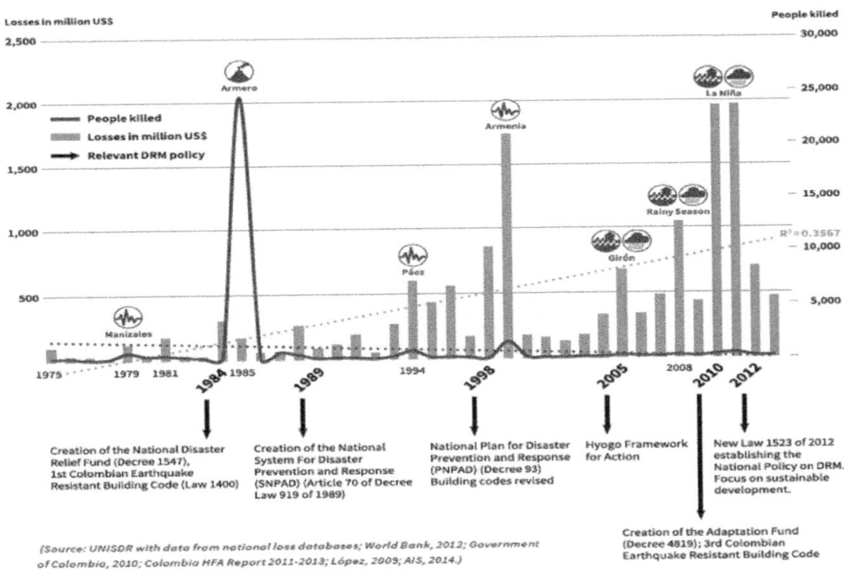

Figure 1.6. *"Disaster risk reduction and loss trends in Colombia (in constant 2012 $US)" from [UNI 15b]. The multiplication of disaster prevention policies in Colombia does not lead to an apparent reduction in losses recorded*

1.2. Disaster prevention policies are proliferating on every scale

At the national scale, as Figure 1.6 shows, we are seeing an effective multiplication of laws and legal standards concerning disaster prevention. To clearly show that this trend is valid despite the economic, cultural and political differences between countries, here we compare France and Japan. This type of approach has already been conducted, for example, between France and a Latin American country, Colombia [LÓP 11, REB 13b]. This normative demultiplication results in institutions that are charged, directly or indirectly, to implement the policies in order to prevent disasters. The appearance of transversal institutions contributes, in return, to increasing the number of institutions involved in disaster prevention even more. It raises question of coordination between these multiple policies and the institutions that run them, whether they are national or international. The underlying trend has also been identified at European and worldwide scales. The frameworks for action already discussed, such as the Hyogo Framework for Action (HFA) and the Sendai Framework for Action (SFA), specifically allow for this. However, this multiplication of policies, frameworks for action and institutions is not preventing the statistically noticeable increase in the number of disasters.

1.2.1. *At the national scale, there is an increasing number of laws and legal standards dedicated to disaster prevention*

The Tohoku disaster in 2011 drew attention to the results of prevention policies in Japan, and there are several resulting summaries, especially in Mochizuki [MOC 14]. Certainly, disaster prevention policies in Japan existed before 1961, as books on the land occupations of the country recall, such as that of Berque [BER 82]. They mention, for example, the regulations and prohibition of deforestation in order to prevent disasters traditionally associated with floods and landslides. All the same, according to Mochizuki, the Disaster Countermeasures Basic Act of 1961 is fundamental in that it includes not less than 107 articles on prevention. That did not prevent the measure from being revised following the Kobe disaster: 16 special new laws strengthening crisis management, between 1995 and 2000. The 2011 disaster justifies reconsidering this legal mechanism again, with a flowering of new laws, that Mochizuki takes pains to break down: not less than 7 laws strengthening aid to victims, 5 for rehabilitation and Machizukuri (strengthening disaster prevention from communities of inhabitants), 3 dedicated to the return to normal for businesses, 6 supporting the local public authorities, 9 related to the nuclear component of the disaster and, finally, 12 others focus, in particular, on the financial aspects of the reconstruction. And again, this inventory, "*à la Prévert*", stops here at no less than 42 new laws but is in no way exhaustive.

It might be thought that this legal inflation is the expression of a national particularity. This might be the case for Japan, which, according to Kelman and Glantz [KEL 15a, p. 105], is in the dominant position worldwide when it comes to disaster prevention: "Japan rose to the occasion, yet again leading the world in this realm". In fact, each major disaster prompts reconsideration of the existing measures based on experience, since, simply, every major event shows the limits of prevention policies that already exist.

Similar trends exist outside Japan, for example, in France. The legal mechanism concerning "natural catastrophes" dates back at least to Tazieff's law, in 1982. It can be seen as an equivalent of the Japanese law of 1961. Likewise, the main French laws for restoring of mountain lands (in French "restauration des terrains en montagne" or RTM) are similar to Japanese regulations, since they are also aimed at preventing disasters associated with deforestation. The RTM laws date back to the years 1860 and 1882. They show that disaster prevention is no more a novelty in France than in Japan. Their age in no way prevents any legal revisions. The 1982 law has been revised numerous times, notably in 1995 ("Barnier's Law"), 2003 ("Bachelot's Law") as well as 2010. The laws are accompanied by their regulations, such as the decree of 1984 which created a tool since modified several times. It was then called the Exposure Plan for Predictable Natural Risks. Circulars have also contributed to the evolution of prevention policies, such as the 2002 one on the land-use zones associated with dikes [PIG 14]. Additionally, the measure was transformed by taking into account the European directives, such as the Flood Directive, of 2007, transposed into French law. In France, as in Japan, the legal evolution is effectively paced by several major disasters that revealed the limits of preexisting measures. They are clearly identified by the reports of experts who follow disasters. In fact, the law of 2010 corresponds to the double disaster associated with the Xynthia storm and to the floods in Var, France when peaks in mortality and above all peaks in damage were recorded.

1.2.2. *At the national scale, there are an increasing number of institutions dedicated to disaster prevention*

The evolution of legal measures is accompanied more or less directly by new institutions. For example, in the case of Japan, the Reconstruction Agency was created by a special law in February 2012. Officially, it has the goal of planning and coordinating different measures in keeping with the reconstruction following the disaster. It draws attention to the specific problems with coordination between an increasing number of institutions, more or less involved in reconstruction and in disaster prevention. This function is mentioned explicitly on the agency's official site, which is national. It represents the role of "control tower", improving coordination between different institutions and "cooperation with local administrations".

The comparison with France is once again instructive: evolution of institutions there is also partially related to major disasters, as well as with the multiplication of legal standards. For example, the European Center for Prevention and Management of Flood Risk (in French: "Centre Européen de Prévention et de gestion des Risques d'Inondation" or CEPRI) is a French organization created in December 1, 2006. It responds to the desire to "reduce the gap between local structures and the State with regard to prevention and management of flood risk. In the past, the majority of communities worked separately, without sharing their onsite experiences. In addition, the State had problems collecting information about what was happening on the ground and sometimes implemented rules that were not adapted to local realities"[1]. This conflict between institutions and the explosion of information had to be revised with the appearance of new management tools, including the Action Programs for Flood Prevention (in French: "Programmes d'Actions de Prevention des Inondations" or PAPI). As CEPRI's official site clearly shows: "a local community or group (department council, township committee, mixed syndicate, etc.) makes up the pilot structure of the PAPI. This group contracts with the State and local actors as part of a multiannual convention that determines the methods of implementation of the PAPI".

Yet, we can see the evolution of the system. The first-generation PAPI were created in 2003, in response to disasters linked to the 1999 (Aude) or 2002 (Somme) floods. Between 2003 and 2009, more than 50 PAPI were implemented all over the French territory. However, owing to the transposition of the European "flood" directive into French law, and above all the major disasters of 2010, the measure was reviewed by the circular of May 12, 2011 relating to the labeling and monitoring of the projects. This trend accompanies the recent multiplication of intercommunal structures

1 http://www.cepri.net/presentation.html, accessed January 14, 2016

being linked to river contracts. This is the case of the intercommunal syndicate of Arve and surroundings, in Haute-Savoie, created in 1988 and holder of the Arve-Léman PAPI. Other institutions, such as the Public Territorial Establishment of Seine-Amont, which manages the PAPI of the same name, are tied to other types of measures taken in relation to disaster prevention, in this case, the reservoirs in the upstream portion of the drainage basin of the Seine. These intercommunal structures attempt to compensate for the institutional fragmentation that reduces the expected effectiveness of policies aimed at preventing disasters.

This multiplication of both standards and institutions therefore poses problems of coordination. Hence, there was appearance of transversal institutions, which are also justified by the necessity of preventing disasters by integrating the supra-local scale. It can be regional, but it often turns out to be national and international.

1.2.3. *More disaster prevention frameworks for action at the international scale*

The 2015–2030 framework for the reduction of disaster risk was adopted by representatives of 187 UN Member States on March 18, 2015 during the Third UN World Conference on Disaster Risk Reduction in Sendai. Paragraph 14 of its preamble states that: "In order to reduce disaster risk, there is a need to address existing challenges and prepare for future ones by focusing on: monitoring, assessing and understanding disaster risk and sharing such information and how it is created; strengthening disaster risk governance and coordination across relevant institutions and sectors and the full and meaningful participation of relevant stakeholders at appropriate levels" [UNI 15a, p. 11]. The first point is concerned with research, but it is presented as being closely linked to the second point, which is aimed at managers and prevention policies. In fact, these links between management and research are confirmed by the sources of the main databases on disasters, as we saw above.

There are still more, with the recent appearance of institutions, at the international scale, that draw attention to the necessity for preventing disasters with more integrated management of policies as well as information.

Integrated Research on Disaster Risk (IRDR) is an applied research project that dates back to 2008. In that year, the International Council for Science, the International Council for Social Sciences and the International Strategy for Disaster Prevention decided to finance it. Indeed, according to Cutter *et al.* [CUT 13], "government agencies, private companies and other organizations may collect and manage data related to their own areas of interest using their own standards and procedures, without significant collaboration with other groups. This results in gaps in the data, inconsistent overlaps, and biases that ultimately affect the quality of research conducted and policies made based on the data".

As expected, the project involves a specific action: a database of the losses resulting from disasters (disaster loss data). It has been led by a federal structure, a "management community". Its goal is to have the academic and management participants collaborated on the identification and implementation of methodological recommendations concerning the collection, storage and dissemination of information about disaster loss. "It plays the role of a forum offering guidelines for the growing network of stakeholders from different disciplines and sectors to study issues related to the collection, storage and dissemination of disaster loss data".

Here we find also on the international scale a "desired integration of science into practice and policy making" concerning disaster prevention [GAL 15]. Nevertheless, these efforts do not attain the desired results, as the same authors lament. In addition, their multiplication seems to justify even more the pessimistic results of prevention policies: officially, the more we manage, the more we lose.

With such a vicious circle, it is therefore no longer surprising to see how much research on disaster prevention is increasing, whether it is applied or not.

1.3. Research on disasters and their prevention is also increasing

The increase in research on disasters and their prevention reinforces again the paradoxical character of disasters, which seems stand out in statistical terms. The natural or inevitable nature of disasters and the portion of the responsibility that human societies bear are questions that have been asked for several centuries. However, those questions still pose a problem today. The research tends to recognize even more the amount of responsibility that human societies have in the creation of disasters, as the links with the theme of climate change recall. Recognizing responsibility also means recognizing the possibilities for action, and therefore, simultaneously justifies the double necessity of knowledge and prevention policies. The approaches mobilized to understand disasters and their prevention evolve and are recomposed as well. This also helps to understand why making global assessments is even more difficult today. All of this is accompanied by a growing profusion of journals and publications, as well as research projects on the theme. The comparison with the statistical evolution of disasters, which seem to be increasing, leaves us thinking that, effectively, the more we know, the more we lose.

1.3.1. *Diversification of schools of thought on disaster prevention*

Controversies about the role of human society in the creation of disasters may be dated, but they remain topical, since the debate is not clear-cut. We verified it concerning efforts at distinguishing between *dis-astrum* and *kata-strophe* [UNI 15b]. It is doubtful that a consensual solution can be reached, even if the statistically significant increase in disasters necessarily calls for an effort for reflecting on prevention policies, and requires, as much from an ethical point of view as an economic policy, clarifying this issue as much as possible. The repeated invoking of the famous debate between Rousseau and Voltaire about the Lisbon disaster in 1,755 simply confirms that the debate is as old as it is topical. The exchange between Rousseau and Voltaire is brought up again today to bring into consideration the question of human responsibility. A famous work by Beck on "the risk society", appeared in 1986, is therefore put in historical perspective vis-à-vis the Lisbon controversy in Portal [POR 09], by the contribution of François Walter: "On the Natural Catastrophe to the Modern Conception of Risk and the Idea of Crisis". The "paradigm of the society of risk" is exposed by Beck himself [BEC 86]: "Therefore it is no longer, or not solely to make nature profitable, to liberate man from traditional constraints, but also and above all to resolve the problems induced by techno-economic development itself. The modernization process becomes 'reflexive', it is itself an object of reflection and problems". Accordingly, also, "Nature, which was a fact, has now become a construct" [BEC 86, p. 16].

Under these conditions, research on disasters cannot only involve the factors that are identified first in case of events, whatever they are. Disaster prevention cannot be reduced to just identifying places where hazardous substances are stored, or even the characteristics of ships transporting hydrocarbons. It also cannot be reduced to geotechnical research on landslides, or hydraulic research on inland floods. These contributions are surely necessary and useful, if limited. In addition, the links between different types of disasters are recognized, for example, with NATECH's expression of disasters. Nevertheless, the separation between the types of disasters is still very clearly topical. The presentation of themes of the journal *International Journal of Emergency Management* shows it very clearly, among others, in 2015:

"Topics covered include: Natural disasters: hurricanes, flooding, earthquakes; Industrial hazards: nuclear accidents, etc.; Transport accidents: road, maritime, air and rail"[2].

It is the main reason for the research on the immediate impact of disasters, which makes up an even more dominant side of research today, and is pursued generally by engineering, as corrective works. Other types of research that focus more on the social and political aspect itself pursue and complete the impact-oriented research. The disaster associated with Katrina would not have existed without the hurricane that gave it its name. However, the interpretation of a disaster makes it possible to verify once again how it is explained first and foremost by a set of elements of vulnerability that the structure of land occupations of New Orleans had, as well as political choices. In this field, we find a body of research on disaster prevention that has been strongly influenced by anthropology, history, political science and, also necessarily, sociology. The expression of "social construction of risks and disasters" [GAR 05] very clearly refers to this part of the research on disasters.

The distinction between the two main approaches, which remains topical, can be easily seen in the structure and content of research on disasters. It will be discussed at length in Chapter 7. The more engineering, geotechnical, geological, geophysical and climatological components are marked by the

2 http://www.inderscience.com/jhome.php?jcode=ijem, accessed January 14, 2016.

hard sciences as much as by earth sciences. Moreover, it includes, even today, a large part of the research on climate change. The plans of academic works on disasters are often marked by very analytical and compartmentalized approaches. They use the classifications depending on the hazards that make up the most immediate and directly visible causes of disasters. By contrast, the social component is focused on a reading that is more evolutionist, constructivist, and in line with the socioeconomic, political and historical characteristics of societies. The international research moreover evokes here the root causes and the underlying risk factors borrowed from the works of radical geography, and summarized in the work of Blaikie *et al.* [BLA 94] republished with Wisner *et al.* [WIS 04]. The theme of research on adaptation to climate change is also subject to this distinction in approaches, while the engineering approach, focused on impacts and disturbances, dominates here as well [BAS 13].

From a constructivist perspective, nevertheless, societies ultimately produce disasters, and have produced them, partly inadvertently, throughout their history. This type of interpretation inevitably calls into question the forms of political action, which have become as useful as they prove to be necessarily limited. Beck's reflexivity is in agreement with this reading.

Of course, recognizing the hybrid character of "nature" transformed and integrated by human societies – as exemplified by this book (see Figure 8.10)– also warrants an increase in environmental research from a socio-ecological perspective on disaster prevention. Several groups of researchers support this transversal and integrated reading, especially starting in the 1970s regarding the group Resilience Alliance [GUN 02]. This was not the only group developing research over the past few decades, but they are accompanied by similar, complementary approaches. We can mention here researchers incorporated into the society IDRiM, created to promote integrated and systemic approaches in the field of disaster prevention [SHI 13] or even the contributions of networks of researchers on socio-ecological systems. Moreover, they can sometimes use very little of the work of the Resilience Alliance, as we can verify with Renaud *et al.* [REN 13].

These groups of researchers regularly publish their work, resulting in an inevitable increase in the number of international journals and articles concerning disaster prevention. In addition, these journals clearly show a desire to link research and management as closely as possible, such as in the monitoring of prevention policies. For example, the journal *International*

Journal of Emergency Management immediately notes, in a banner on its website, the necessity of strong links between research and management: "linking academia, business and industry through research".

1.3.2. *Evolution and diversification of international journals on disaster prevention*

It is not difficult to verify the recent multiplication of journals that are concerned with the theme. On September 8, 2015, we simply cross-referenced two sources of information on the Internet. One of the sources corresponds to the results of a search carried out with the Google search engine, using entry, "*Journals on Disaster Management*". The other source comes from the information provided by a site dedicated to the same research, Environmental XPRT[3], and which appears, moreover, on the first page of Google results. This site intends to facilitate access to disaster management magazines and journals. The existence of this site, on its own, proves how difficult it is to establish an inventory today, and therefore, inevitably, the multiplication of journals and articles on this problem.

We note that the two sources, nevertheless necessarily very close, provide results that overlap very little as a whole. The Environmental XPRT site provides 12 entries, and the first page of the Google search provides 10 results. If we simply cross-reference the two sources, we find no less than 5 international journals appearing on the first page of the Google search that are not mentioned by the research site Environmental XPRT. This very preliminary approach to the literature, which has the advantage of being simple to carry out, confirms the recent proliferation of journals on this theme. Their number obviously exceeds by far the very reduced sample that we have here. For example, journals such as *Global Environmental Change*, *Natural Hazards* or even *Disasters*, not to mention journals in languages other than English, such as *VertigO*, are simply ignored by the Environmental XPRT site. This reflects criteria for selection that are not mentioned by either of the two sites but which also draw attention to the market of access to information. Finally, it is revealing to note that none of the journals that appeared in the initial approach to the

3 http://www.environmental-expert.com/magazines/international-journal-of-global-environmental-issues-ijgenvl-3464, consulted January 14, 2016

two sites dated from before 1992. Six of them, almost half of this reduced sample, appeared in 2009 or after.

In fact, a part of the research on disaster prevention is now also interested in bibliometrics. McEntire [MC 07, p. 297] approaches the question from the information sciences, and over a short period. A similar action was conducted by Gall *et al.* [GAL 15]. The article takes up a part of the work that the IRDR research group conducts in order to determine the level of integration (an Assessment of Integrated Research on Disaster Risk), that is to say, information sharing and collaboration between multiple institutions and groups of researchers who work on disaster prevention. "During 2014, the working group completed its bibliometric analysis of English language peer-reviewed scientific journals published between 1999 and 2013, as well as 78 other English language publications from the period 1993 to 2011" [IRD 15, p. 7]. We do not need to detail the results of this research here. Just its existence confirms that the multiplication of journals and articles on disaster prevention poses problems of readability and accessibility as well as knowledge. It is in line with the editorial objectives that the *International Journal of Climate Change Strategies and Management* has: "In the history of science there have been only a few issues which have mobilized the attention of scientists and policy-makers alike as the issue of climate change currently does. The *International Journal of Climate Change Strategies and Management* addresses the need for disseminating scholarly research, projects and other initiatives, such as new policies, strategies or action plans, which may lead to a better understanding of the subject matter of climate change"[4].

Indeed, as reflected by editorial initiatives, research projects on this matter are multiplying.

1.3.3. *Evolution and diversification of international research projects on disasters*

Today, we can clearly observe an inflation of international research projects on disasters. The European Union officially supports research

4 http://www.emeraldgrouppublishing.com/products/journals/journals.htm?id=ijccsm, accessed January 14, 2016.

projects that contribute to a better understanding and improvement of the results of disaster prevention policies. We can read on the website of the *Official Journal of the European Union* that, "the 7th framework program of research and development (RDFP) supports the coordination of research activities and the distribution of information concerning the risk of catastrophe"[5]. In fact, research projects on disaster prevention are already old, as numerous and difficult to count as the policies they are attached to. The authors of this book were able to participate in several of these projects, whether as part of the European cooperation responsible for humanitarian aid and emergency preparedness, with ECHO [DER 99], or in that of the RDFP programs. Some of them have given rise to the publication of books: that of Menoni and Margottini [MEN 11] uses in fact the results of the European program SCENARIO. It must be emphasized that these programs are not only focused on Europe. The KNOW4DRR project included a comparative component on Vietnam and Mexico [SPI 15]. The results of this project have also been able to contribute to identifying where the principal problems encountered by disaster prevention at the global scale are, in line with the efforts at reflection that justify the Sendai International Framework for Action [WEI 15].

We can therefore make an initial assessment from these analyses, which conclude: the paradox of the gap between losses, research and prevention policies would have intensified since the work of White *et al.* [WHI 01]. At least for a portion of disasters, the multiplication of databases and policies dedicated to prevention or international research is fundamentally called into question when we read what the International Council for Science (ICSU) page says about the project that motivates the IDRiM initiative: "The impacts of natural hazards continue to increase around the world; the frequency of recorded disasters affecting communities has risen significantly over the past century. Hundreds of thousands of people are killed and millions injured, affected or displaced each year because of disasters, and the amount of property damage has been doubling every 7 years on average over the past 40 years"[6]. The paradox is found regarding the theme of sustainable development: never has it been so much of a question as at a time when disasters are officially increasing [PIG 07]. Never has the expression

5 http://eur-lex.europa.eu/legal-content/FR/TXT/?uri=URISERV:pr0005.
6 http://www.icsu.org/what-we-do/interdisciplinary-bodies/irdr, accessed January 14, 2016.

"sustainable development" been so close to an oxymoron [DAH 13]. Yet, it is precisely the principle of sustainable development that the International Council for Science refers to in order to justify support for the IRDR project. The latter is part of the "ICSU Priority Area Assessment on Environment and its Relation to Sustainable Development (2003)"[7].

Despite the contradictory messages, it seems difficult to take into account an evaluation that is so paradoxical. This needs to be reconsidered, which is what we propose to do in the second chapter of this book.

7 http://www.icsu.org/what-we-do/interdisciplinary-bodies/irdr, accessed January 14, 2016.

2

The Contributions to Disaster Prevention are Difficult to Assess

Reports officially presenting disasters and their development are surprising in their overwhelmingly dominant pessimism. Indeed, there is no lack of evidence attesting that prevention policies do their job, even if they encounter numerous limits. In this second chapter, as a counterpoint to the first chapter, we will present arguments that put the pessimistic, but dominant, messages regarding prevention policies in perspective and add nuance. In the first place, not all of the statistical databases on disasters show an increase in frequency or even an increase in the average intensity of the associated damages (2.1).

Moreover, despite the limits and methodological difficulties of comparison, it is simplistic and questionable to consider the evolution of disasters, on the one side, and economic and demographic changes, on the other side, separately and especially exclusively. Sectoral studies are admittedly useful and aim for the highest possible precision. They certainly have numerous limits, which Hallegatte and Théry [HAL 07, p. 508] mention, regarding adaptation to climate change: "Even more striking, these studies are all founded on a limited group of sectoral studies, and their authors are the first to recognize the limits of these evaluations and to place a low level of confidence in them". Segmenting studies on disaster prevention in the extreme do not therefore lead to a gain in accuracy. However, an approach only centered on the study of disasters inherently favors a first impression that is pessimistic, simply because it only focuses on losses. Conversely, comparing losses and economic (2.2) or demographic (2.3) issues, even in the form of an order of magnitude, prompts the relativization of these same losses.

We do not encourage considering economic losses and especially human losses, as ultimately secondary, nor even less accepting them. Quite the contrary, we want to attract attention to the possibility of limiting losses, even if there are always too many of them. It is reasonable to better understand why this possibility of limiting losses is given so little attention in the reflection on prevention policies, as well as in the reports provided. It also helps us to see how we can hope to limit these losses even more. Even if, as we will see, following the literature, it is utopian to claim to be able to eliminate all losses. This suggests an additional element, which should necessarily encourage reconsideration of the ordinarily sobering reports on prevention policies.

Indeed, the possibility of limiting losses also explains that disaster prevention, as such, makes up the markets (2.4). Disaster prevention concerns economic sectors such as building and construction, technical and legal consulting and insurance. Disaster prevention is an approach that is as necessary from an ethical perspective as it can be profitable on the economic front. This relationship between prevention policies and economic actors has been mentioned multiple times, whatever the countries and cultural areas concerned. We even find it associated with the history of capitalism.

All of these elements justify yet more asking of why, under such conditions, such a pessimistic presentation of reports on disaster prevention still dominates today.

2.1. Disasters are not increasing according to all databases

A counter-example is enough to verify this point. This is what we propose to offer in the field of disasters associated with marine hydrocarbon pollution, as well as with oil spills, which are sometimes in the news. Although oil spills are not all caused by maritime transportation of oil in tankers, since pollution can also come from off-shore drilling platforms, the relationship between transport and pollution is sufficiently recognized to justify several statistical databases on this type of disaster. The subject has also produced academic works, such as, in France, those of Paul Fattal [FAT 06] and Julien Gardaix [GAR 09]. This is not surprising since France has been affected by several oil spills, such as the one following the wreck of the

Erika in 1999. International and national recognition of disasters linked to maritime hydrocarbons pollution gave rise to developments very similar to those which were described in the first chapter: specific databases, particular prevention policies and dedicated new institutions that accompany the development of research and specialized journals.

The messages that the different databases bring to light are contrasting, at least in part, because the available statistics show a reduction in certain types of disasters. This change can be verified by cross-referencing two independent sources, one international, the ITOPF database, and the other French, the Center for Documentation, Research and Experimentations on Accidental Water Pollution (CEDRE).[1] It turns out that the ITOPF database concerns major disasters while the CEDRE database also includes disasters with a lower intensity of damage. For disasters called "technological disasters", we find subdivisions between types of disasters and related databases that are reminiscent of the distinction between EM-DAT and DesInventar. At first glance, dividing the types of disasters into technological and "natural", seems to be supported by the evidence, that is, the different elements involved. However, this separation masks characteristics common between types of disasters, more fundamental, which the differences between the databases and above all their complementarities reveal. Jérôme Domenichini's [DOM 08] thesis has the distinction of drawing attention to this point. It specifies and discusses the similarities, seen by insurers, between disasters linked to floods and those that are associated with terrorism.

The ITOPF database of disasters associated with maritime transport of hydrocarbons was created after the wreck of the *Torrey Canyon* in 1967 off the coast of Great Britain. It released more than 110,000 tons of crude oil into the ocean. This ship was one of the first super-tankers. International feeling, and doubtlessly the fact that the British coast was affected, drew a reaction from ship owners. They then developed the Tanker Owners Voluntary Agreement Concerning Liability for Oil Pollution [TOV 69], with an associated organization, the ITOPF. The institution had the goal of managing the aftermath of oil spills as well as contributing to the prevention of this type of disaster. Such an approach also involves developing knowledge of this type of disaster using a specific database.

1 http://wwz.cedre.fr/, consulted January 14, 2016.

The database clearly shows a drop in the frequency of major disasters, as a screen capture of ITOPF's official site (Figure 2.1) demonstrates. It also shows a drop in the average intensity of damage, assuming that the latter is simply a function of estimated volume of the pollution-causing hydrocarbons that are released into the ocean. This is of course not the case: research studies show the multifactorial nature of pollution and the associated damage [FAT 06]. The damage associated with oil spills varies, according to, among other factors, the characteristics of the pollutant, the navigational conditions and the structure of the ship as well as depending on whether the decision to tow is made. This reservation immediately draws attention to the information's limits. Moreover, some of these limits are not unknown to the ITOPF site itself. Furthermore, and by definition, the site only includes accidental pollution. Nevertheless, deliberate pollution related to maritime transport exists. This pollution is not necessarily illegal, as the first article of the United Nations Convention on the Law of the Sea, created in 1982, known as the Montego Bay Convention, and its paragraphs 4 and 5 specify. Obviously, this type of pollution is not included in the ITOPF database. The drop in frequency that the ITOPF database identifies therefore concerns only major disasters, unintentional ones. In addition, this drop is more than probable as major disasters are also the most visible. The formula, according to which the more we know and the more we manage, the more we lose, must therefore be partially reconsidered. Notably, this more optimistic report is partially confirmed by other sources such as the CEDRE database.

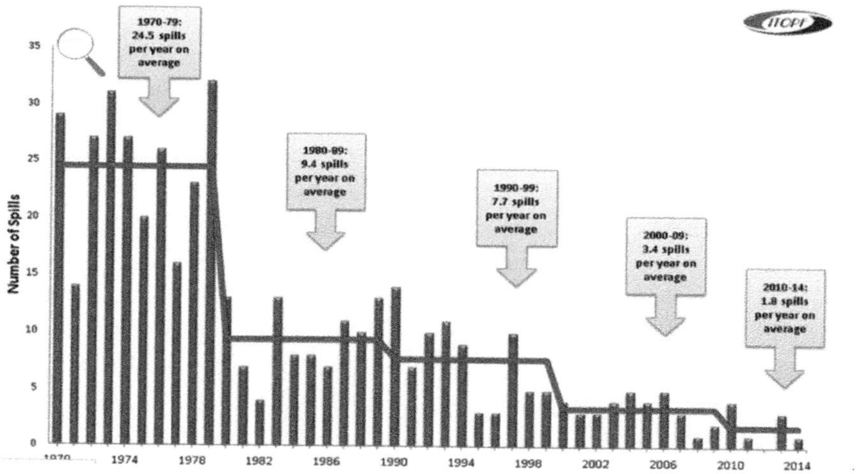

Figure 2.1. *Evolution of the frequency of disasters associated with oil spills according to ITOPF [ITO 15]*

In France, there is another database on disasters associated with maritime hydrocarbon pollution, the CEDRE database. As with the ITOPF database, CEDRE was created following a major disaster, the one that accompanied the wreck and oil spill of the *Amoco Cadiz* in March 1978. This disaster also mobilized international public opinion and national public opinion, this time French. As it explicitly mentions on its official site, "CEDRE capitalizes on knowledge and is continually developing tools to fulfil its different missions". The institution, created in January 1979, also maintains a database on "accidents at sea", which overlaps with information from the ITOPF database. However, CEDRE also collects official reports on pollution (POLREP). This type of information, which is more recent and was especially strengthened following the wreck of the *Erika*, in 1999, also concerns pollution with a lower level of damage, and which is intentional. The information is determined by the particular legal criteria related to traffic in the most monitored seas, such as the English Channel. Maritime surveillance areas were established there based on the legal framework that the 1982 Montego Bay Convention on the Law of the Sea facilitates.

The statistical database collecting POLREP information shows a relatively regular growth in the frequency of pollution since 1990, the first year for which usable statistics are available according to CEDRE. However, as the official CEDRE site indicates, "this growth could be more of an appearance than a reality during the 1990s, since the statistics from the beginning of the decade seem to be marred by a number of omissions". Recall that these statistics do not include legal pollution and that some of the events with a more moderate intensity of damage are not counted. We could therefore suppose that low-intensity pollution and small-scale disasters were increasing in frequency, despite the monitoring and increased penalties, as suggested in Figure 2.2. Conversely, higher intensity pollution, in the form of major disasters that are obviously the most conspicuous, is seeing its frequency decline.

We can see how difficult interpreting these statistics is, something that leads to the close critical debates found in particular in the works of Fattal [FAT 06] and Gardaix [GAR 09]. The fact that we can infer three key elements from the reports on the type of disasters remains. The frequency of major disasters, and of course associated unintentional pollution, is decreasing. It is likely that the frequency of lower-intensity pollution is increasing, especially the part of this pollution that is legal and therefore invisible in the statistics. This double development does not preclude the possibility of a major disaster occurring over the upcoming years and decades. We know *a posteriori* that even ships recognized as being among the most reliable can nevertheless sink, as the example of *Derbyshire* shows. This ship, which was constructed in 1976, was classified A1 by Lloyd's Register, which is the insurance company's best possible grade. Nevertheless, it sank with its crew of 42 men in 1980 [ISE 14].

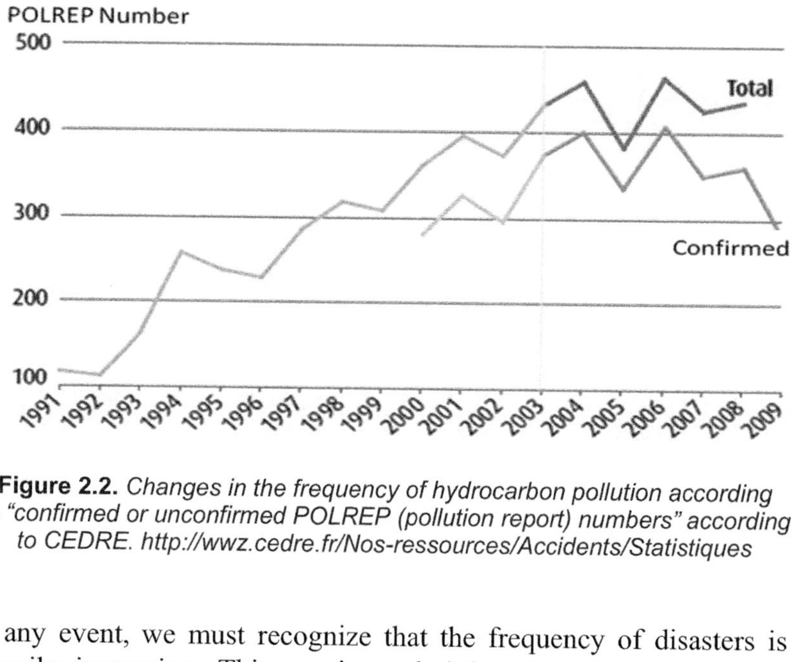

Figure 2.2. *Changes in the frequency of hydrocarbon pollution according to "confirmed or unconfirmed POLREP (pollution report) numbers" according to CEDRE. http://wwz.cedre.fr/Nos-ressources/Accidents/Statistiques*

In any event, we must recognize that the frequency of disasters is not necessarily increasing. This requires admitting that the multiplication of information, institutions, supervision and policies contribute at least in part to the reduction in the number of disasters. This is of course the case for those disasters associated with the most intense damage. For small-scale disasters, we can also ask what the situation would be in the absence of political regulations. In fact, the evaluation is missing what cannot be measured, namely the disasters that have been avoided, since by definition no statistical database exists for them. In brief, the Montego Bay Convention on the International Law of the Sea, as well as the reactions of some of the stakeholders in the sector of maritime transport of hydrocarbons, is indispensable to understanding the reports, even those that are partial and therefore tricky to interpret, that are available to us today [GAR 09]. For example, the desire to review the structures of ships by developing double hulls can show not only the capacity for adaptation, but also adaptation's limits.

This kind of development immediately demonstrates the economic as well as ethical issues linked to disaster prevention. It also encourages a reconsideration of the paradox of losses accrued despite the multiplication of research and prevention policies.

2.2. Not necessarily more disasters, but definitely more to lose

Indeed, if we can discuss, on the one hand, the increase in the frequency of small-scale disasters concerning maritime pollution during the transport of hydrocarbons, on the other hand, the increase in the risk of this type of disaster, including major disasters, is easy to show. Despite the larger methodological difficulties, a similar message is received once we compare the macroeconomic and demographic developments and those concerning changes in the frequency and intensity of the damage associated with disasters, whatever be their type.

Returning to disasters associated with maritime transport of hydrocarbons, the ISEMAR [ISE 14] note mentions the trend in "internationalization" and "complexification" of the routes taken by tankers. Some of these ships are also found on routes that are particularly exposed. This is what the traffic on the Northeast Passage, which passes to the north of Russia, shows, even though it remains limited. The navigational conditions there are particularly difficult even with the technological support of the Russian icebreakers. Furthermore, quantitatively, the statistics on the international transport of hydrocarbons show a doubling in refined hydrocarbons transported between 1999 and 2012, and an increase of at least 10% in the transport of crude oil over the same period [ISE 14]. Thus, it is clear that there is a rising trend in major disaster risk tied to international maritime transport of hydrocarbons. We can compare this trend with the statistics despite the limits already addressed. It is therefore barely possible to understand the drop in the frequency of major disasters without integrating into the reasoning the capacity for management, namely, the forms of political intervention and the capacity for reaction that the research could support. We therefore do not necessarily lose everything by knowing and managing more.

Moreover, losing more while managing more assumes that the reports are integrated. However, whatever the types of disasters, the statistics do not necessarily reflect the evolution of losses in relation to that of the goods exposed, and even more to the issues at stake, far from it. The issues at stake define "what can be lost" [MET 11]. Above all, they correspond to the desire to find the comparatively greater vulnerable critical elements from among multiple elements, here, the goods exposed. In fact, the most accessible statistics only concern losses and direct losses. For example, the statistics from Swiss-Re and Munich-Re show the direct losses concerning insured or non-insured goods. Information on the markets of these businesses, namely, on the goods insured and the amounts of the premiums collected, remains more difficult to find. The comparison between the two elements of the report remains more confidential. We propose to verify this very segmented and focused type of approach on losses (as well as on hazards) with the report for the year 2003 that Swiss-Re provided in its online archives [SWI 03]:

"Five insured losses in excess of one billion US dollars in North America and none in Europe or Asia – this is the preliminary conclusion drawn for 2003 by Swiss Re's Sigma. Although, at USD 17 billion, losses were higher than in the previous year, 2003 is still no record year in a long-term comparison. Sigma's records indicate that there were a high number of fatalities: almost 20,000 people lost their lives in the just under 350 registered catastrophes. More than 2,200 people perished in the earthquake in Algeria in May, while nearly 1,400 died in India's heat wave."

According to the first part of this official commentary, 2003 was not a record year in terms of economic losses. However, the reinsurance company presents the report by stressing the human losses. This suggests and seems to explain that losses for which insurance companies must compensate are increasing despite everything. According to the extremely ambiguous commentary, the year 2003 is not a year associated with record losses if the length of time is taken into account, but it still records more losses than 2002. The ambiguity is found in the title of the report itself. It explains the conclusion of the report and the following:

"Burden for property insurers: rising trend.

Insured property losses from natural catastrophes and man-made disasters in 2003 amounted to an overall approximate USD 17 billion. Although this figure is below those of recent record years – an inflation-adjusted USD 36 billion (including Hurricane Andrew) in 1992, USD 35 billion (including terrorist attack of 11 September) in 2001, USD 34 billion (including storms Lothar and Martin) in 1999 – 2003 can be added to the list of record losses. Natural catastrophes have caused billions of losses in virtually every year since the late eighties (an inflation-adjusted annual average of USD 20 billion since 1987). The 2003 figures confirm this trend towards high losses, which is being driven in part by increasingly densely populated areas, higher concentrations of insured values and the development of endangered zones."

As can be seen, the commentary on the report from the reinsurance company only provides information about the losses, and it stresses the trend in the increase of the latter. Despite the beginning of the commentary, 2003 actually became a year of record losses. For good measure, the company also displays a trend graph whose goal is to visually confirm the written commentary, and which seems to be authoritative (Figure 2.3).

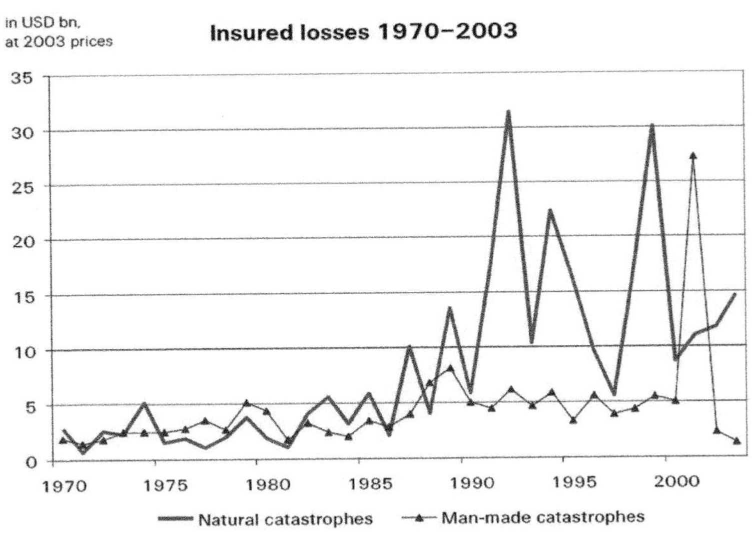

Figure 2.3. *The evolution of losses recorded by SwissRe between 1970 and 2003, in SwissRe [SWI 03]. http://media.swissre.com/documents/pr_20031216_sigma_en.pdf*

Curiously, the outline of the explanation concerning the increase in losses: increase in population density, densification of goods insured, does not cause the commentary to put the losses in perspective in relation to the changes in goods or lives exposed and to the amounts of the premiums that insurance companies collect. Yet, if all of the activities, goods and persons are far from being insured, the losses should be compared, among others, to the amounts of premiums collected. This comparative approach, which seems merely numerical and simple, obviously exists, but we find it more readily in other aspects of the insurance field. For example, it explains an indicator used for comparing European countries with each other, namely the insurance penetration rate: "The penetration rate measures the percentage of dwellings which are insured against a given NatCat or, in other words, the percentage of global insurance premiums over a country's gross domestic product" [MAC 12, p. 17]. Furthermore, according to Michel-Kerjean [MIC 13, p. 617], "The field of the economy of catastrophes is booming". We can find arguments in terms of evaluations in thesis works, such as Domenichini [DOM 08]. In Figure 2.4, the rate of damage of the insured sums therefore compares the insured goods with the losses registered by the insurance companies, in this case, when the Rhône flooded in Arles, in 2003. The insured sum, or the insured capital, is "the amount set in the insurance policy that becomes due when the risk insured occurs". The lower the percentage, the more the losses are limited for the insurance company compared with the premiums collected, for which we still do not have firsthand information here. For Domenichini [DOM 08], the goal is elsewhere. The data from certain insurance companies, which are not directly accessible, makes it possible to simply and clearly show that the damage cannot only be explained by the height the water reached during the flood. This is moreover not valid, whatever the type of damage in connection with insured or uninsured goods, and even with regard to mortality, which we will return to below. The comparison between losses and goods exposed is therefore mobilizable and sometimes mobilized either at the macro- or micro-scale. However, this comparative approach remains partial, indirect and discrete in the most accessible reports.

Yet, even if it involves rough orders of magnitude that would be in no way acceptable as such in a work of research on economics, it is clear that the macroeconomic estimates of production and transportation activities cannot be expressed in the same order of magnitude as those concerning the loss of insured goods. The losses of the insurers are assessed in billions of dollars, but the macroeconomic evaluations of only the international exchanges of commodities, for example, reach values estimated in thousands of billions of dollars. This is the information that the statistics of the United Nations Conference on Trade and Development [UNC 15, UNC 14] provides. According to them, international commodity exchanges only were estimated at a little more than 2000 billion dollars in 1980 and at more than 18,800 billion dollars in 2013 [UNC 14, p. 22]. The estimated development of worldwide trade in service is of an even higher order of magnitude: it was estimated at 395 billion dollars in 1980 and at 4,720 billion dollars in 2013 [UNC 15, p. 260]. We mention these orders of magnitude to draw attention to the potential for growth in the insurance business and the collection of associated premiums. There is no simple and direct relationship between the increase in the estimated value of the goods in circulation and the evolution of the premiums that the insurance companies collect: they vary, in particular, depending on the penetration rate. Moreover, the UNCTAD [UNC 14] also evaluated "the evolution of exports and imports of insurance services": 27 billion dollars in 2000 and 102 billion dollars in 2013 [UNC 14, p. 258].

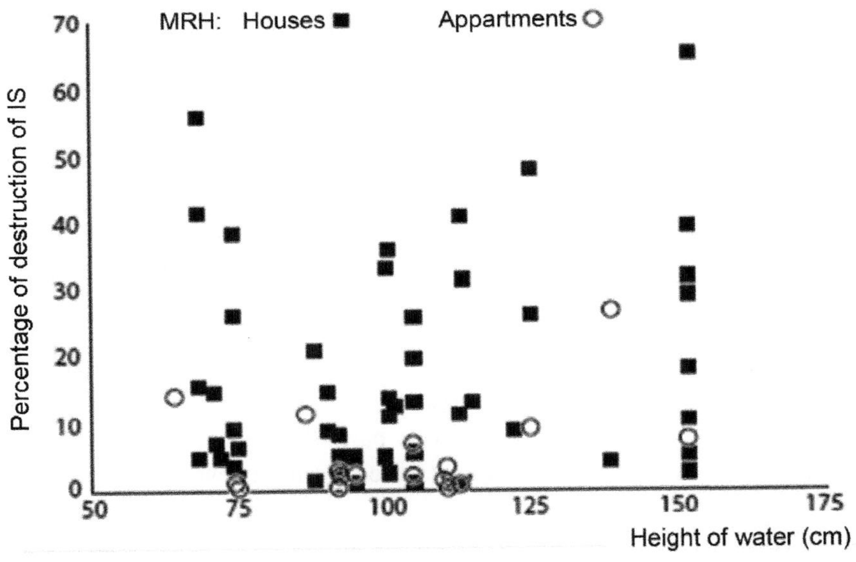

Figure 2.4. *"Absence of a direct relationship between recorded damage and depth of inundation during the flooding in Arles in 2003". From Domenichini [DOM 08, p. 75]. IS insured sums (sums insured); MRH: multi-risk home. Indirect use of the ratio between losses recorded and sums insured*

Even if these are only statistical orders of magnitude, comparing them with the estimated changes in losses for insurance and reinsurance companies considerably weakens the paradox according to which the more we would manage, the more we would know, and more would be lost. Simply put, if we lose more in absolute value, it is also that there is more and more to lose, and there is above all more to gain, even if that is not true for everyone. This statement can seem simplistic, but the fact remains that it is formulated under the pen of some economists themselves. Here, beyond mathematical formalism, Zajdenweber [ZAJ 01] shows an essential part of the problem: "The amount of the damage depends fundamentally on the wealth of a country and on the concentration of this wealth in denser and denser agglomerations". This does not prevent the same author from writing in the same article: "All of these dramatic events disturb insurers since, never in the history of insurance have catastrophes with a human origin affecting developed countries (outside of wars and the large urban fires of the past, such as the London fire of 1666) reached the amount of damage comparable to that of the most serious natural catastrophes, climatic or telluric. A new source of extreme events therefore needs to be taken into account in evaluating these events and the corresponding assessments."

Moreover, the interest in this comparative approach has been recognized for a long time, along with the numerous methodological obstacles that are associated with it. We will not develop this point, nevertheless as essential as it is sensible, any further. We refer here to previous works, such as that of Albouy [ALB 02]. On page 18, he mentions that: "A final perversity is to only consider the negative effects of human activity on natural spaces and the environment and to never practice double-entry bookkeeping of what technology and progress make possible to preserve it and to make it habitable". Assuming that economic losses are increasing, at the same time there has never been so much to lose, and therefore so much to gain, for certain actors. Nevertheless, for reasons partially linked to methodological problems, "We know very little about the economic consequences of natural catastrophes, even if they happen regularly and provide us with a quantity of data" [HAL 07].

Here, we find the academic and institutional tendency to underestimate disaster prevention by putting more emphasis on the estimation of losses reached in absolute value. It is also clearly identifiable regarding mortality.

2.3. There is not always more mortality associated with disasters

The changes in mortality associated with disasters, whatever be their categories, is similar to that involving disasters associated with maritime transportation of hydrocarbons. Certain databases, such as EM-DAT, show a recent trend of a drop in worldwide mortality, and this is even in absolute value. Already in 2002, Albouy was noting on page 13 that "Catastrophes have become one of the minor causes of mortality". However, he also mentions that "This very low mortality from catastrophes masks the real impact of the phenomena" [ALB 02, p. 35], regarding, this time, the people affected.

This statistical drop in mortality would therefore be even more significant in relative value, and we should be glad, even if mortality is still too present. We recall, here again by relying on international statistics, that the world's population was estimated at 3 billion in 1960, 5.3 billion in 1990 and around 7.3 billion in the summer of 2015. In fact, certain countries, such as Colombia, Japan and Spain, clearly show a slight absolute drop and an even larger relative drop in mortality associated with disasters referred to as "natural" [OLC 16]. These trends are repeated in the GAR Report [UNI 15b]. For France, Laurent Boissier's thesis [BOI 13] clearly shows an absolute downward trend and an even larger downward trend in mortality, if assessed in relative value, associated with floods in the Mediterranean Midi. Here again, as with major disasters associated with maritime transportation of hydrocarbons, we should reconsider the statement according to which the more we know and the more we manage, the more we lose.

However, as with evaluations of economic losses, critical approaches to the information mitigate the scope of these reports. They are much more difficult to establish than they might seem at first. The subject is taboo, and it is difficult to find comparable sources of information. The question of division between the multiple factors which explain the mortality also proves that direct causality is very difficult to demonstrate despite initial appearances. For example, mortality during heatwaves involves as much as it reveals first of all the most vulnerable. Heatwaves hastened their deaths: epidemiological studies and the geography of health confirm this. According to Besancenot [BES 04], during the heatwave in 2003 in France, "Seven out of ten of the supernumerary deaths were not the direct result of a failure of the thermoregulatory mechanisms. Heat can also aggravate a preexisting illness or contribute to triggering it. First in these causes of over-represented deaths during the heat wave were pathologies of the genitourinary system (+240% in total) and illnesses of the respiratory system (+180%)". We often see dips in mortality after disasters associated with heatwaves, a phenomenon identified in Brussels in 1994, according to Besancenot, even if this does not seem to have been the case in France in 2003. These elements immediately attract attention, once again, in the absence of a simple and linear causal relationship between disaster and mortality, and even between hazard and intensity of the damage associated with disasters. The heatwave clearly appeared to be the immediate cause of mortality, but it masked numerous other factors that predisposed it more fundamentally. Finally, in which statistical categories should these deaths be placed? What are the limits of categories to be considered, and based on what criteria?

Even more important, mortality reports such as those issued by EM-DAT are also questioned by UNISDR [UNI 15b]. According to this institution, the EM-DAT database does not count "small-scale disasters" and "extensive risks", as we mentioned above. In fact, at the global scale, it is evident how much mortality figures can differ even when the characteristics of exposure to a hazard might be similar. That was the case during the passage of Cyclone Coleen, which affected Vietnam, Northern Cambodia and the western part of Thailand consecutively in October 1992. The losses declared in Thailand were higher compared with Cambodia, with the cyclone having lost its intensity by moved onto the land, and even though it involved land occupations with similar characteristics in the two countries [DER 99]. These discrepancies reflect at least in part, the differences in the collection of information, and in access to the latter. They suggest a hidden mortality. If a human being is poor, or a member of an ethnic minority, she or he is less visible in the statistics. These differences also justify the interest in distinguishing exposure from vulnerability.

Finally, we find several very interesting arguments on the changes in mortality associated with disasters in the work of Mitchell *et al.* [MIT 14]. The authors emphasize the very large heterogeneity of the statistics coming from sources that are themselves varied. They describe it as "noisy". Even more importantly, on page X of the executive summary, they state that, "It is not possible to establish a true statistical average for mortality or economic losses from only a few decades of national loss data. An example of this is in Haiti, where earthquakes killed fewer than 10 people between 1900 and 2009 before killing over 220,000 people in a single afternoon in 2010". As a result, any report made from the existing statistics can provide only limited information. On page IX of the same summary, we find, "Accordingly, until it is possible to produce a reliable global assessment of the risk of losses across a range of hazards at country level, the establishment of targets around disaster losses is as much an art as it is a science". Significantly, Mitchell *et al.* [MIT 14] end by providing a trend graph concerning worldwide mortality, and it shows a slight increase, in absolute value, of the latter (Figure 2.5).

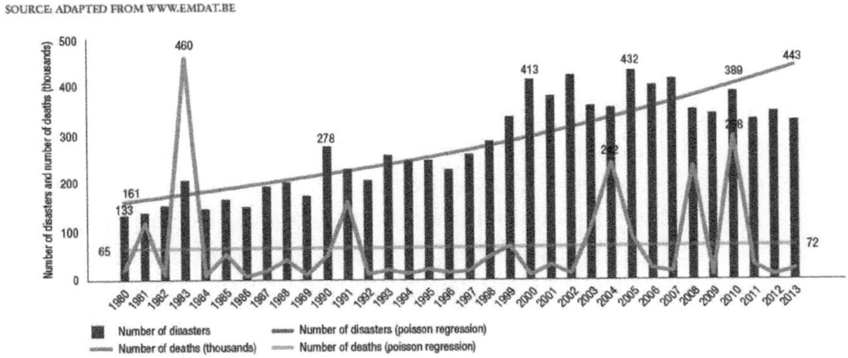

Figure 2.5. *"Global trends in disaster events and death tolls, 1980–2013". In Mitchell et al. [MIT 14, p. 6]*

Consequently, the problem of obtaining a report that is as precise as possible leads us to question the reality of this visible statistical drop in mortality, sometimes in absolute value, and mostly in relative value. However, reconsidering the statistical mortality reports, such as those that are provided by the EM-DAT database (with a downward trend), also calls disaster policies into question. Indeed, the surprising drop in mortality recorded in the EM-DAT database, at a time when humanity has never been so numerous, could also reflect an at least partial contribution of prevention policies to this reported trend. It would remain to demonstrate this hypothesis, although it is discussed in several research studies on the regional [BOI 13] or national [OLC 16] scales. Whatever be the methodological difficulties encountered in this type of approach, the participation of CRED's director, Debby Sapir, in critical works by Mitchell *et al.* [MIT 14], is enough to acknowledge the limits of the EM-DAT database that CRED manages. The EM-DAT database's imperfections would in no way allow it to serve as a benchmark for any disaster prevention policy. Critical readings of databases on mortality therefore draw less attention to the possibility of preventing disasters than to the limits of policies and research. Yet, prevention exists, even if there cannot be a simple causal relationship between prevention policy and statistical changes in mortality.

Overall, several points can be deduced from this critical and necessarily polyphonic reading, necessarily very debated, which involves the lessons of statistics on the frequency of disasters, as well as the recorded economic and human losses that are associated with them.

First of all, the frequency and the losses should be systematically presented with respect to the issues at stake, and their development, which is not yet necessarily the case. The comparative approach, despite its limits, or even its rough nature, usefully draws attention to the evaluations that are not necessarily as negative as their first impression suggests, overall. We cannot completely rejoice that the losses must be relatively lower, and even less the human losses. One death is always one death too many. Furthermore, long-term awareness should prompt more restraint when it comes to optimism.

Next, the interpretations related to disaster statistics, despite their limits, necessarily indicate the existence of prevention policies. It is hard to understand why the losses would be so relatively limited, even over a relatively short period of time, while the issues at stake were increasing so much, if the prevention effort did not exist. When recorded, the absolute statistical increase in the losses therefore indicates limits of prevention policies more than their absence or, even less, their total ineffectiveness. The question could be reformulated another way: what would the human toll be for countries such as Bangladesh in the absence of prevention policies such as those that exist regarding cyclone warnings and civilian safety measures, for example [GLA 09, GLA 15]?

Under these conditions, we are entitled to question the paradox that states that the more we know and the more we act, the more we lose. We have seen that this paradox is defensible, at least at first glance. However, it loses some of its relevance on a second reading. We can therefore legitimately ask ourselves why it has obtained such international recognition, almost unanimous, and even inspired research projects. In fact, it represents the academic side of the pessimism that the institutions in charge of disaster prevention readily display. This almost generalized pessimism, though already criticized, clearly poses a problem with its apparently unequivocal aspect. It is even more the case as soon as we recognize how much disaster prevention can also shape markets.

2.4. The markets interested in disaster prevention

Disaster prevention is not only justified on the ethical front: it also makes up and structures numerous markets. Therefore, the question draws attention to its affinities with the different forms of capitalism. It has given rise to numerous research studies and syntheses, principally in history or in economics. Urban geography also helps to identify more directly the elective affinities between disaster, the development of metropolitan functions and the sometimes very quick developments, that make up the metropolitan landscapes.

The markets in disaster prevention are visible on multiple fronts. Possibly, in the first place due to the desire to control the physical processes, such as landslides or avalanches that contribute (at least partially) to explaining disasters. If humanity does not have the means to act directly on the focus of an earthquake, the geophysical mechanisms of a volcanic eruption or on the formation or the path of a cyclone, it can, on the other hand, act at a more local scale. This can be done by intervening in both the physical processes, using corrective works, and in the characteristics of land occupation, such as by adapting buildings, and/or by evacuating the population. For earthquakes, there are rules for earthquake-proof construction. For certain volcanic eruptions, it takes the form of constructing anti-lahar works, which the mega-dike of Pinatubo shows [GAI 08], or, among other things, aimed at diverting basalt flows. For cyclones, which are accompanied by wind and precipitation, there are architectural adaptations such as dikes and other hydraulic structures. The scenarios for evacuating the population are definable [ROB 09] and they are implemented, as during the eruption of Merapi in 2010 [PIC 13].

Management of civil safety, that is also supported by the consultancy firms of design offices, is and has to be made visible by sign panels, such as in Aramon, in the Gard [PIG 10].

All of these elements are now standard and give rise to a very extensive literature. It is enough here to refer to the example of the dams known as "sabo" used for correcting torrents, particularly in Japanese mountainous landscapes. They are not only found near Japanese volcanoes, but also modify the flow of watercourses that drain off the Merapi Volcano, in Indonesia, where they contributed to the management of the eruption of 2010, as well as the associated lahars [DE 12, PIC 13]. Similar torrent-control corrective works are identifiable in countries as culturally and physically different as France and Bolivia. They assume today both the intervention of consultancy firms, often foreign, and the companies in building and construction. However, they are not totally new. France has often used alluvial dams as part of the policies known as restoration of mountainous lands, starting in 1860, especially in rural environments. We should not think that this construction of corrective works has had no economic implications. Consulting the archives [PIG 91] shows that building alluvial dams, for example, in the commune of Vailly in Haute Savoie provided salaried work to farm workers during the winter. Moreover, in 1940, the goal was also to combat local unemployment and to get the demobilized male populations to work. Bolivia continues to favor these solutions, known as geotechnical solutions, to prevent disasters in the La Paz Basin, as Sébastien Hardy's atlas illustrated and showed [HAR 13].

In France as in Bolivia today, torrent-control corrective works, whose stated goal is to prevent disasters, result in a part from the markets involving building and construction companies. Augendre [AUG 08] also mentions the importance of their lobby in Japan, and their links with political parties even if this type of practice is not necessarily generalized. Concerning Japan, Marie Augendre argues that "everything comes together on the concrete path" [AUG 08, p. 164] and that "building and construction in Japan today" still seem to be "angel investors". She notes the difference in Switzerland, where "the cost of public works does not exceed 3.5% of the GDP, while in Japan, it uses twice that" [AUG 08, p. 166]. Therefore, it should not be surprising to see works, even obsolete ones, still integrated into land occupations (Figure 2.6). Such an economic and geographic importance cannot be explained only by the goal of protection, but also encounters limits. In fact, "construction firms are closely connected to the political class. If we remember that the accounting of businesses has for a long time remained opaque, we can understand that these have been easily able to support the highly costly political activity of the Liberal Democratic Party (LDP), which is in power" in Japan [AUG 08, p. 166].

Figure 2.6. *Importance of torrent-control corrective works such as "sabo" dams (Kobe, Mt. Rokko, August 2014)*

The burden of building and construction and costly engineering responses can also be seen within the range of risk management strategies in Ecuador. In 2014, there was a national survey on the risk management strategies of 61 out of the 221 local governments making up the country [REB 15]. The 61 local governments had undertaken 76 risk management projects (completed or in progress), since 2009 in collaboration with public institutions and with international cooperation, for an overall budget of more than 46 million dollars. From studying of the content of these programs, it can be seen that close to 44 million dollars, which is 95% of the financing of the projects indicated by the local communities surveyed, involved the construction of protective works or interventions on the natural environment. In this situation, since 2008, a national institution (the SGR) has guided the decentralization of risk management. This institution's budget was raised to several dozens of millions of dollars in 2014, while the organization that was most involved in risk management before 2008, Civil Protection, had several million dollars only. In the middle of the decentralization process, the local communities, which are consolidating little by little and assuming new functions, are applying for national financing not only from the SGR, but also from the Banco del Estado or from different ministries in order to develop their local risk management strategies, very marked by technological solutions and concrete. Moreover they solicit financing offered by international cooperation, and use the services of dozens of NGOs and aid agencies (such as the Spanish AECID or the German GTZ) as part of a large "market" whose fragmentation (concerning financing as well as the nature of the services offered) is not without its own problems for coordinating a national risk management strategy.

Another major actor in prevention is represented by insurance companies. The basis of this activity rests on the mutualization of risk. It involves sharing insurable risk between multiple partners forming a market. The mutualization of risk is particularly present in France, where insurance and mutual companies are institutionally associated with the FFSA-GEMA group. For example, according to Azema [AZE 07], "Groupama is a mutualist company; we are born from local initiatives and a community of interest – originally farmers coming from the same area who share the desire to secure their holdings against risk". The insurance business is associated with several principles whose goal is guaranteeing coverage but also profitability. This is the case with the franchise system: its goal is to make the activity profitable by eliminating damage with high frequency and relatively low intensity (Figure 1.2). For this category of events, the intensity of the damage remains comparatively limited, if it is considered in isolation, but its high frequency makes insurance activity unprofitable. It even calls into question its basic principle, which is to cover risk in case damage occurs.

Finally, it is useful to remember that risk management and disaster prevention are primarily urban, and especially metropolitan, affairs.

Fundamentally, the economist Zajdenweber [ZAJ 01], among others who are not economists but mathematicians [BAK 99], draws attention to the fact that towns and disasters demonstrate similar statistical characteristics. Zipf's Law applies to networks of towns, by verifying, despite differences, that their frequency is the inverse of their demographic weight. However, Zipf's Law belongs to the same group of power laws as Farmer's: as we have seen, making it possible to distinguish between small-scale disasters and major disasters, to the point where it can provide a supporting graphic (Figure 1.2) to explain the differences in definition between *dis-aster* and *kata-strophe* [UNI 15b]. The Farmer Curve is also useful for justifying where insurance can intervene (Figure 1.2).

On their own, these statistical similarities are not enough to demonstrate the existence of favored functional relationships, and even less explanatory ones, between towns, disaster prevention, and the insurance field: they do suggest it as Zajdenweber also does [ZAJ 01]. Mathematicians consider power laws as a means of identifying complexity [BAK 99], but the latter does not limit evidence to the three elements taken into account, it exceeds them by a great deal. It is also used in the field of ecology, among others [GUN 02] for subjects that, on the surface, do not directly concern towns or disaster prevention. This is the case for changes in coral reefs or in the swamps of the Everglades. Effectively, the history of ecology [DRO 93] makes very clear the urban and even metropolitan origin of the discipline, as for every scholarly discipline. The history of ecology also shows its close association with the desire to find solutions to the problems identified as such by metropolitan institutions. Complexity is certainly not reducible to the understanding of relationships between towns and disaster prevention alone. However, ecology and its historical developments suggest that complexity has been mobilized more specifically in the hopes of responding to needs identified as priorities, and potentially disastrous, by certain metropolitan institutions.

The history of towns, insurance and capitalism makes it possible to confirm, *a posteriori*, and above all very explicitly, their partial association. Disaster prevention effectively represents at least potential markets, and it is linked to the history of metropolises, especially European ones [BRA 85].

Insurance and reinsurance companies were born in the urban, and especially metropolitan, environment as developed by, among others, Chaline and Dubois-Maury [CHA 94]. Here we can see simply the fact that towns favor, potentially, disasters by the concentration of issues at stake, by the different forms of mobility that the exchanges require. But they equally favor, and also potentiate the countermeasures that the disaster prevention policies represent [PIG 12a]. Towns are by definition the most common forms of land occupation that prompt the setting up of policies for preventing disasters. The recognition of their limits encourages pushing back against or bypassing them. Disaster prevention is therefore one of the major engines of urbanization, where it is at least as much by its limits as by its success, always partial.

The concept of risk was born in the port towns of Medieval Italy, where accounting and banking innovations, such as double-entry bookkeeping, were also found [BAE 95, BAI 85, BEN 93]. Risk is therefore first and foremost a metropolitan issue. It emphasizes the possibility of reducing and taking advantage of the marine risk, which accompanies the metropolitan management of international or global-scale exchanges, without being able to eliminate all disasters. This possibility for action also requires the most independent forms of decision making possible, without referring to an external and dominant power. A metropolis necessarily confirms itself as such against the other, by its normative power, for example by its outreach not only in the field of the economy. It is therefore not surprising that theorists of capitalism, such as Schumpeter [SCH 42], or historians of capitalism, such as Braudel [BRA 85], have also showed these elective affinities between towns, disaster prevention and risks, where the potential was finally increased. Schumpeter notes that "capitalism, we repeat, constitutes, by its very nature, a type or a method of economic transformation and, not only is it never stationary, but it can never become so". Moreover, the exact formulation that made this author famous refers to "the perennial gale of Creative Destruction" [SCH 42, p. 106] very significantly.

These elective affinities can be observed. One of the modern London's most striking buildings, the Gherkin, is sometimes still called the Swiss-Re building, since the reinsurance company built it and occupied it between 2004 and 2007. It continued the much older effort of disaster prevention, which justified the formation of Lloyd's. It was and remains essential to the British metropolization, to the rapid expansion of trade and shipping lines. The Gherkin Building also occupies a site where the Baltic Exchange Building was found, dating from 1903. The latter was damaged in 1992 in an attack committed by a branch of the IRA. It was an attack on one of the most notorious elements of London's power. The Gherkin building is therefore symbolic on many levels. It shows the metropolitan capacity to transform disaster prevention into very beneficial economic activities. It also incorporates several forms of capitalism, as much tied to trade and insurance as to speculations on land values. This is reflected moreover by the forms of urban landscapes marked by renewal, the unexpected and disruptions.

Overall, we hope that Chapter 2 has demonstrated both why we can be surprised to find such a pessimistic interpretation of disaster prevention policies, and why this pessimistic reading can, at least in part, persist.

Indeed, the dominant pessimistic reading underestimates or even masks the human opportunity, if not necessity, to prevent disasters. Certain databases on disasters, such as the ITOPF database, bring to light a downward trend in the frequency of major disasters, and others, such as

EM-DAT, a drop in recorded mortality. At first glance, this should be good news. Furthermore, the increase in losses should never be dissociated from the increase in issues at stake, and it is this that makes it possible to understand why markets in disaster prevention, such as those of insurance, can exist. We can therefore further question the orthodoxy claiming that we always lose more by knowing and managing more.

However, the pessimistic evaluations can be supported by another critical rereading. It highlights the very real underestimation of small-scale disasters, or mortality figures concerning the poorest populations, which are also very often minority populations as well. The same interpretations also claim that the sources of statistical information are too imperfect, and, finally, that useful and reliable knowledge about disasters remains too limited.

It does not prevent our reflection from confirming a real anomaly: that which consists of favoring a pessimistic approach concerning disaster prevention policies. It seems to be justified by the numerous limits that the knowledge about disasters encounters as much as the prevention policies themselves. But we can assume that these limits are not the only possible explanations, far from it. This encourages us to look at those limits more closely.

PART 2

Understanding and Managing Disaster Risk: The Multiple Limits of Specialized Approaches

The second part of this book is dedicated to the usual and predominant interpretations of the paradox according to which the more we know and the more we manage, the more we lose. We should therefore understand better in order to manage better in the hopes of losing less. The limits of policies primarily reflect the limits of our knowledge of disasters: this supposition explains why research programs on disaster prevention have multiplied. The research is no longer as much on disasters as on the disaster risk. It is a matter of looking further toward prevention by relying on what was produced earlier, in improving the understanding of disasters and, above all, what causes them. The first part of the book confirmed how national or international research programs on disasters, and especially their prevention, are multiplying.

The research on disasters and disaster risk in the earth sciences, and also in economics, sociology and geography, has at least one point in common. It involves the makeup of new databases mentioning both the limits as well as the deficiencies of databases that predate the research. According to academic disciplines, these databases are the foundation of efforts at interpretation that can mobilize several forms of quantification and cartography. According to Propeck-Zimmermann [PRO 15a, p. 33], "analysis, evaluation and management are three terms that [...] include different dimensions in the process of apprehension and processing of risk". However, there are steps in order to hope to "prevent and control risk situations".

3

The Limits of Understanding Disasters

We come back, therefore, to the databases, the essential components of research efforts. Indeed, they make up the foundations of the knowledge that we can have about disaster risk. The expression of the database, as such, also suggests that they play a fundamental role in analyzing problems, which are here linked to the development of disasters and the associated damage. The very principle of the databases is by no means called into question as much by academic institutions as by managers involved in disaster prevention. A forum was held in Antalya in February 2015, which brought together several international institutions directly involved in disaster prevention, such as the World Meteorological Organization, the World Bank and the International Red Cross, besides several dozen researchers from more than 30 countries involved in international research programs on the same question. The Antalya Forum prepared, among other things, for the Sendai congress of March 2015, in order to review the principal framework for international action dedicated to disaster prevention. Yet, the first recommendation resulting from a week of collective reflection in Antalya, and whose concluding statement suggested six initiatives for action, displays the necessity of using a new database. Accessible by a portal, they rest on post-disaster returns on experience and on what we can learn from them.

"Six Calls-to-Action

1. 'Lessons Learning' Process: The Need for a 'Lessons Identified' Portal

A user-friendly, innovative DRR knowledge portal should be established to focus specifically on collecting, verifying, cataloguing, archiving, transferring and sharing both positive and negative DRR-related lessons identified during previous DRR-related interventions. These lessons can then serve as the bases for more efficient and effective future interventions"[1].

The participants hope that this database will make it possible to avoid losing information concerning disasters and their prevention. This information is useful for constructing new forms of shared and discussed knowledge when political decisions concerning disaster prevention are made.

Nevertheless, the limits of databases and of what we can expect from them are also recognized today. The expression of data is obviously misleading; a piece of data does not have value in absolute terms. It necessarily assumes reconsidering a preexisting evaluation, and it incorporates multiple presentations. A piece of data is everything except random. Moreover, the still historically recent recognition of the concept of uncertainty [GOD 02] reveals the limits of quantification, and quantified data, which does not mean that the effort to measure or collect the information have lost all interest (section 3.1). The limits of databases also draw attention to two other elements that are just as fundamental. The definitions of risks continue to pose multiple problems despite the recent developments. This situation is extremely surprising and frankly unexpected in view of the issues that are attached to the question of disasters and their prevention. It confirms and reinforces that which we have already seen regarding definitions of disasters. How can databases on disaster risk be structured in order to refine the existing information, while the basic definitions of risk pose so many problems, just like those of disasters themselves (section 3.2)?

This is even more surprising due to the extreme difficulty of finding an interpretative model concerning disaster risk that has unanimous acceptance in the research (section 3.3). The limits that academic research on disasters and their prevention reveals seem to effectively prepare those that the management side of the question encounters. The fourth chapter of this book is dedicated to them. It is particularly marked by the recognition of policies that remain very segmented, according to the multiple institutions and actors that are involved in disaster prevention.

1 http://www.ccb-boulder.org/wp-content/uploads/2015/03/English-Antalya-Statement.pdf, consulted January 16, 2016.

3.1. The place and status of inherent uncertainty in databases

We have already presented *supra* several major databases, including the EM-DAT database and DesInventar. The limits that these databases show do not concern the aspects that have already been addressed, such as those that make the distinction between direct or indirect damage, or the establishment of criteria making it possible to include, or not, the deaths listed in the databases. The limits of access to information, or even of its composition (or not), that has already been addressed *supra*, and that determine the databases on disasters, is no longer enough to exhaust this subject. Nevertheless, they had already drawn attention to the fact that a database, whichever it is, is nothing but just a moment in the process aimed at refining knowledge of disasters and their prevention. It is essential to mention the many-sided conditioning tied to this specific moment. A database cannot be neutral, or apply in absolute terms, contrary to what the expression would suggest.

This part is more specifically dedicated to underlying problems posed by the comparability of the data that make up, once collected and assembled, databases on disasters. It addresses equally the use and the scope of the results that allow obtaining a database for the purposes of knowledge and even of prediction. In fact, the beginning of this chapter engages in a critical reflection on a concept that we have not yet addressed in detail, which is the concept of risk. Indeed, prevention policies are now officially concerned more with the reduction of disaster risk than just on disaster prevention as such. An entire field of research is dedicated to disaster risk reduction. Nevertheless, there is no unanimity about the concept of risk itself. The problems that comparability of data poses, such as the use of information from databases for the purposes of reports or even prediction, make it possible to understand why the concept of risk tends to be replaced by that of uncertainty. It is therefore not that surprising that risk and its definitions pose so many problems today. We propose to justify this development on the basis of the results of an international program in which one of the two authors participated directly.

Between 2005 and 2008, a program for promoting international intra-European cooperation, the Franco-Swiss INTERREG, was developed on the Haut-Rhône, concerning risks associated with water supply and with inland floods [DUP 08]. For the second component dedicated to floods, the approach, very traditional and expected, if not more necessary, involved collecting historical data. It demands identifying the sites that are the most exposed to inland floods, and those which are susceptible to being even more so today taking into account current developments. It is also involved in identifying trends in the preparation of disasters. A database on the damage linked to floods was therefore created by critical cross-referencing of the different sources of historical and current information available.

The result very clearly shows a downward trend in the frequency of harmful inland floods in the Arve Valley. This result is consistent with the information collected on other parts of the drainage basin of the Rhône, for example, where the floods are more monitored, such as in Lyon. Claire Combe's thesis [COM 07], together with that of Emeline Comby [COM 15], draws attention to the fact that the decrease in the frequency of harmful floods should not encourage us to think that any future disaster has been prevented. Assuming this requires being able to show that the prevention policies carried out, in particular with inland river corrective works, would make it possible to eliminate any future damage. This is not the case, since, simply, the very conception of any corrective work, such as a dike, acknowledges its limits. A dike is officially sized depending on the estimated statistical return of a flood. Whatever this return period is, the work is necessarily limited; in fact, it is about specifying what these limits can be. This conception of corrective works by engineers on its own draws attention to the interest of these works for events whose occurrence presents a high frequency, and their limits for events with a low frequency. It is also necessary to be careful of the significance of the period of time included in this statistical analysis. By definition, an event with a very rare frequency appears rarely on frequency histograms concerning a limited period.

The reading of the "sleeping city" [COM 07], lulled by the effectiveness of its dikes against events that would have a high frequency in the absence of a dike, is just as defensible in the Arve Valley. Recall that it includes part of the Geneva urban area. Efforts to defend as much in Geneva as in Lyon have not stopped increasing despite the explicit recognition of the limits of the corrective works by the engineers who design them. It is therefore as legitimate as it is useful to identify areas of the Haut-Rhône drainage basin whose developments in land occupation contribute the most to future disasters. To do this, it is necessary to construct a database, compare the data among itself, highlighting trends in the frequency of events, and also map the results.

However, once presented this underlying trend of inadvertent contribution to disasters, above all owing to the limits of corrective works that are ultimately taken into account very little by the managers of urbanization, what exactly is the value of the comparison between 1800 and 2005? And what can we learn from this database? Indeed, both the climatic conditions and the flow conditions of the Arve have been completely modified between 1800 and 2005. For the literature, 1800 is still in the Little Ice Age. The data on the flood of 1800 do not necessarily have the same meaning as that of 1988. In 1800, historical maps described a runoff that is still woven and multichannel throughout the middle part of the Arve Valley, between Cluses and Bonneville, or even downstream of this town. The Arve again flooded a large part of its flood plain, every year according to the archives, which Mougin's [MOU 14] reference work summarizes. This type of runoff has almost disappeared today. This can be explained by the increased diking, but also by the drop in the coarse load that the current Arve transports with regard to what it transported at the beginning of the 19th Century. This development is not only linked to climate. The work of restoration of mountainous lands, developed after the annexation of Savoy by France, in 1860, contributes to trapping the coarse load of the torrents and mountainous rivers upstream of the drainage basin, among other factors identified. Disaster prevention associated with torrentiality seems therefore to have borne fruit. It seems, at first, to give credit to an expression that became established, and famous, that of the "extinction of torrents" [MOU 14].

However, at the same time, the Arve disperses less of its energy by transporting the reduced coarse load. A part of this energy is now dissipated alongside its riverbed. The Arve cuts through numerous segments of its course at the expense of superficial formations, rocks from the Quaternary Period and weaker cohesion than the older rocks, known as bedrock. This development was obviously not desired by the managers of works such as bridges and even dikes, whose foundations are scoured, and which it was decided to protect against. Field studies often show several generations of corrective works, such as two sills and two related groups of dikes, at the confluence of the Arve and the Foron de Scionzier (Figure 3.1). This development was not anticipated by the planners of policies for restoration of mountainous lands, that were developed starting in 1860. This same development contributes to explaining why overflow floods and the associated damage have become rarer in the database. The underlying trend was to trade, without completely wanting to, higher flood frequencies of the

19th Century, and the associated damage, against the frequency of damage to corrective works and crossings. This damage moreover becomes more and more numerous with time: this is what both the archives and field studies confirm (Figure 3.1). However, the frequency of overflow floods has, effectively, become lower.

Figure 3.1. *Two generations of dikes and sills at the confluence of the Arve and the Foron [DUP 08]*

As a result, the reduction in frequency of harmful floods in this part of the Northern Alps does not mean the disappearance of disasters. It instead means a future event with an admittedly very rare frequency. However, this flood will be much more harmful than the flood plain previously, mainly because agricultural land now has a higher value due to housing developments and zones of economic activity. As we know, a disaster denotes an event that is infrequent but has a comparatively high intensity of damage. Moreover, the identified development further necessitates returning to the corrective works that are present (and damaged) as on the policies carried out.

Can we also use the database to prolong the approach in terms of frequency, by offering probability calculations concerning the statistical population that floods from river overflow represent? It is mathematically possible. However, we know that the result would necessarily be very offset in relation to future events, since the population reference statistic is increasingly heterogeneous as time goes on. Indeed, the more time that passes, the more corrective works and prevention policies increase

(Figure 3.1). The "harmful flood" data inadvertently integrate these multiple parameters that are not immediately visible.

In particular, these limits of databases and of their use for the purposes of applied knowledge and prevention justify the switch from the concept of risk to that of uncertainty. The concept of uncertainty is usable once, according to Godard *et al.* [GOD 02, p. 13], we find "what cannot be surrounded by a distribution of objective probabilities", that is to say, by probabilities taken from the history of past events. Uncertainty is also identified in case of "probabilities deemed not credible" [GOD 02, p. 49]. The authors add that uncertainty is acknowledged in case of "impossibility of establishing probabilities when faced with a new situation", and even faced with the "inability to determine a complete list of the possible results of the action" [GOD 02, p. 38].

These definitions of uncertainty, simultaneously precise and broad, immediately show that the concept appears fundamental once the limits of knowledge as well as management are recognized. These limits are inherent in the databases themselves. In addition, these limits are, in fact, being recognized more and more. An essential part of the research in quantitative economy or in earth sciences is dedicated to the desired identification and reduction of uncertainty. We will illustrate the point with two short examples, one coming from the research, and the other from the world of insurance.

Does volcanic risk really deserve to be called by that name? If we look at the history of eruptions of the Indonesian volcano Merapi, one of the most disaster-prone volcanoes on the planet, it is very quickly seen that the frequency of eruptions remains at less than 10 since the end of the 19th Century ([PIC 13, p. 49], Table 3.1). In concrete terms, these eruptions show characteristics and developments only partly foreseeable by geophysicists, who must update their scenarios according to the events recorded. It is hard to understand what the interest here might be in the probability calculations, while they are nevertheless supposed to establish the concept of risk. It is clearly much more about identifying and hoping to reduce uncertainty. The work of geophysicists like that, complementary, of specialists in vulnerability, contributes to it. They are done in conjunction with policies linked to civil safety [ROB 09] or to land-use zoning, adopted locally and very unequally accepted.

Year of eruption	VEI	Volume of material ejected (millions of m^3)	Deaths	Injured	Evacuated and displaced individuals
1872	4		200		
1904	2		16	45	3,000
1920	2	5	35		1,000
1930	3	26	1,369		13,000
1954	2	20	64	57	3,000
1961	3	29.4	6	6	8,000
1994	2	3.5	64	500	6,026
1998	2			314	6,000
2006	1	1.3	2		22,253
2010	4	**150**	**349**	**1,814**	**1,000,000**

Sources: [VOI 00, LAV 98, WIT 05, MEI 10]

Table 3.1. *History of disasters related to eruptions of Merapi. From [PIC 13]. VEI: volcanic explosivity index*

The shift from risk to uncertainty, which shows the limits of quantification while trying to reduce them, is also identifiable in works connected with insurance companies. This second example concerns the risks linked to storms. Of course, insurers would like to officially have tools that make it possible to refine their knowledge of land occupations and goods insured that are most likely to be damaged, if they were not already using them, at least in part, in an extremely competitive field of activity where information is fundamental. They have therefore asked modelers to provide them with indicators on the sites that are most exposed to storms, hail and snow, particularly the most vulnerable. However, the results that the models provide correspond to only a very small part of the land occupations for which, *a posteriori*, insurers know that the damage is the highest. If we consider Figure 3.2, we can immediately see anomalies appear. Brittany should be more concerned, in view of the damages that the insured people declare following storms that have a high frequency compared with other parts of France, and this is not the case. The problem does not come from the composite character of the types of damage taken into account, assuming that the damage caused by snow or hail, more limited in Britain, would compensate for the even higher damage associated with storms. This discrepancy reveals above all the limits of interpretations that consider the hazard the dominant, if not exclusive, factor in the preparation and the interpretation of disasters. Climatology, which constructs models using the hazard, can therefore only be of partial help in predicting damage and its prevention in this case. It is clear that, just like the case of the floods of the river Arve, the database on the damage integrates many parameters other than those that are only connected to hazards. The discrepancies between that which is expected *a priori* and that which is noted *a posteriori* simply show plurifactoriality. This problem can be found very clearly on the local scale, as once again Camille André's thesis [AND 13] has demonstrated regarding the interpretation of the damage associated with the Xynthia storm, in France.

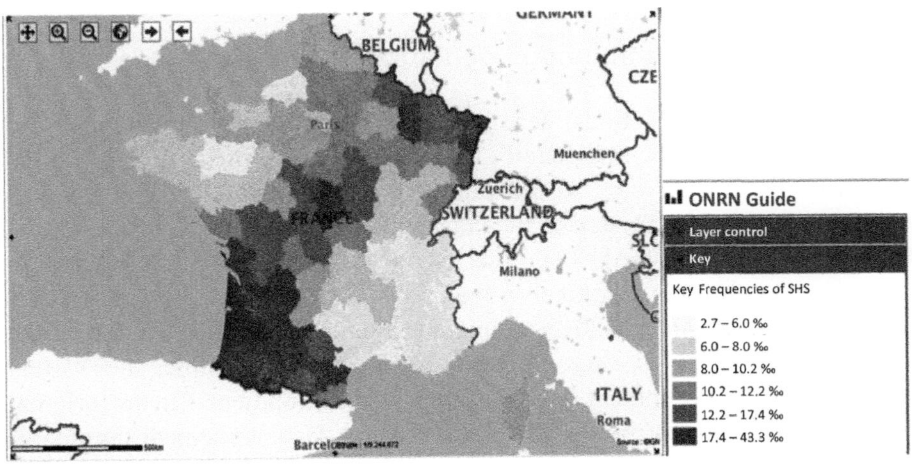

Figure 3.2. *Frequency of "storm-hail-snow losses" according to the ONRN website 2015*

The limits of databases therefore show plurifactoriality, such as the multiple conditionings, methodological or otherwise, which predate and structure them. This is nevertheless what the "data" set up and structured into "bases" hide. These numerous limits are revealed by multiple *a posteriori* observable discrepancies between that which is expected from information taken from databases, and possibly from policies the latter can inspire, and that which is noted when disasters occur. Effectively, research as much as policy therefore gives the impression that the more we know and the more we manage, the more we lose, if we do not take these limits into account. The concept of uncertainty tries to give sense to these discrepancies. It also establishes the possibility of hoping to reduce these discrepancies and limits, without being able to eliminate them. Moreover, the concept of uncertainty reconstructs that, more fundamental, of risk.

Therefore, we should not be too surprised to find many differences with respect to the same definition of what could be a risk. Research is not able, any more than the managers who are even more practically involved, to offer definitions that create a real consensus on this fundamental concept. The concept of risk nevertheless continues to justify structural differences according to the types of databases, in particular according to the multiple disciplinary specificities and divisions, or depending on the type of hazard involved.

3.2. Disaster risk: ambiguous definitions

With some exceptions, the literature is almost unanimous in acknowledging that the risk denotes the probability of damages related to the presence of at least one hazard and exposed elements, whose unequal vulnerability is today related to the concept of issues at stake [BLA 94, PIG 05, BIR06, D'ER 09, DAU 13]. Disaster risk corresponds to the anticipation of damage with a comparatively higher intensity, as we have seen above. The websites available in 2015 on this question simply address a part of the bibliography, which is very voluminous. Take, for example, the Virtual University of Environment and Sustainable Development (in the original French: l'Université Virtuelle Environnement & Développement durable or UVED), an academic site which represents a group of online courses and gathers the contributions of 80 authors, on risks and themes related to management of the environment. Here, we read in a very standard manner that:

"The usual definition given for natural risk is the following:

(Risk) = (Hazard) x (Issue at stake)"[2]

Hazard denotes an "event or process". According to the UN, a hazard is "*a potentially damaging physical event, phenomenon and/or human activity which may cause the loss of life or injury, property damage*" (UNISDR, quoted in Birkmann [BIR 06, p. 15]). The hazard also allows the definition of a frequency, which is a connection between a temporal period and the number of events that occurred during this period. This possibility justifies the use of statistics and probabilities that establish the concept of risk.

2 http://www.e-sige.ensmp.fr/uved/risks/1.1/html/2_2-2_1.html#2-2-2, consulted January 14, 2016.

Indeed, according to the National Center for Textual and Lexical Resources (in the original French: Centre National de Ressources Textuelles et Lexicales or CNTRL) site, the Italian etymology of the concept "*risco*" refers to marine risks, engaging the maritime trade[3]. It also refers to the possibility of running risks and even taking precautions against them. The risk is quantifiable and can be partly anticipated. Here, maritime insurance finds one of its foundations. A low-frequency event and even one with a high intensity of damage can be covered by insurance thanks to the mutualization of losses, the compensation of the insurance premium to be paid. A ship involved in trade activities matches this definition.

Risks can be run, and some even benefit from this. Risk is a financial and insurance-related concept, which lends itself to forms of quantification that are closely linked to several types of capitalisms, as well as to the functioning of metropolises. Indeed, as mentioned *supra*, the risk also supposes the most possible freedom and autonomy in making decisions, as well as in accordance with contracts made [PEY 95, GUE 14b]. But, is this still an approach to risk based on the hazard? The latter is far from enough to account for risk on its own. We can even ask why it is ranked first. An event without possible loss has no meaning here, and that which can be lost necessarily preexists the hazard.

Hence, the issue at stake, the other component of the definition that the UVED site provides, and that it presents are as follows: "The issues at stake and vulnerability are tied to human presence (persons, habitations, economic activities, infrastructure, etc.) and are difficult to define". This is actually the case: hundreds of theses and entire works of summaries have not so far been able to provide a definition that achieves a real consensus. According to Birkmann [BIR 06, p. 12], vulnerability denotes "an intrinsic predisposition to be affected to or to be susceptible to damage" even if "The different definitions or approaches show it is not clear just what 'vulnerability' stands for a scientific concept" [BIR 06, p. 11]. The concept of issues at stake has been redeveloped in the French-language research, in particular by D'Ercole and Metzger [DER 09a]. They maintain that "we can simply say that risk is the possibility of losing that to which we give importance. Yet 'what can be lost' has no conceptual status in the 'hazard x vulnerability paradigm'. The challenge is therefore to provide coherence that is both conceptual and operational to the concept of risk. It is also to consider 'what can be lost,'

3 http://www.cnrtl.fr/etymologie/risk, consulted January 14, 2016.

that is to say the issue at stake, as an autonomous object in the problems of risks, to separate it from hazard and vulnerability that structure the concept of risk to dissociate clearly what can be lost (the issues at stake) from that which can cause their loss (vulnerability)".

As we can see, this position has only been partially adopted by the UVED in the definition of risk that the site provides. This is not surprising. We suggest that the reader consult Birkmann's [BIR 06] summary on the question of vulnerability. It reproduces the multiple efforts at definition. The main question is really to understand why it seems utopian to be able to reach a consensus on such a fundamental question in view of the related issues, as much ethical ones as economic ones. If there is no consensus on defining such a fundamental concept, can we expect more from databases that are necessarily structured according to the varied interpretations of the essentials?

Certain components of the definitions of risk that are now traditional, such as exposure, are not considered by the UVED site. According to Birkmann [BIR 06, p. 23], exposure concerns the "structures, populations and economy" susceptible to being affected by a hazard, while "vulnerability has a physical, social, economic and environmental dimension". According to Shi and Kasperson [SHI 14, p. 3], "exposures mainly include social and economic elements". We see therefore that the definitions of risk vary according to the authors.

Moreover, it should be noted that the UVED site distinguishes, in a very conventional way, between natural and technological risk. It defines the latter this way:

"The engineer characterizes risk as an entity with two dimensions:

(Risk) = (Probability) x (Gravity)"

It is nevertheless clear that the structure of the definition is very close to that which involves "natural risks". The two definitions overlap fundamentally. Here, we find the hazard linked with probability and also vulnerability/issues at stake with gravity. Further proof is provided by a reference to the work of Farmer. The UVED site acknowledges that this work can also include the risks known as "natural risks": "Risk is therefore considered to be a measure of the dangerous situation that results from the confrontation of the hazard and the issues at stake. This measure is often expressed in terms of gravity and probability and, such as for technological risk, can be represented in Farmer's diagram".

Indeed, we can read here the work of this engineer (Farmer) who has worked in the nuclear field, and confirmed that this reading is equally possible in the field of "natural risks". Here, we also find the inverse relationship between frequency (hazards, events) and the intensity of the damage ([PIG 10], Figure 1.2]). It is therefore very surprising to note once again [PIG 94] that the separation between the types of risks and of disasters, whose pertinence we can question, continues to be practiced in the research. It contributes to obscuring the possible definitions of the concept of risk.

This marked reticence is more strongly calling into question the categorization of risks and disasters into types, highlighting another surprising bias. Indeed, referring to the insurance or engineering origins of the concept of risk should draw attention to the place that the definitions devote to the human capacity for action. On principle, an insurer seeks to limit the losses for its own activities (and also for the insured, etc.) by favoring prevention policies. This is even more the case for an engineer who seeks to prevent the damage at least in connection with the work that he conceives and constructs. Yet, it can be seen that the traditional definitions, even revised to include the concept of issues at stake, always place human societies in a position of passive subject. The concept of vulnerability, along with the famous "natural hazards", draw attention to an interpretation that is not very coherent with what any observer can remark, not only *a priori* by a simple effort at reflection, but also *a posteriori*, by going into the field.

Do not think that these elements are linked to an outmoded representation of that question that defining risk poses today. Indeed, the Bureau of Geological and Mining Research (BGMR)[4] site demonstrates a very instructive comparison between the definition of risk that is present and the illustration offered. On the drawing of the assumed cliff, we can now only see a single component of the definition of hazard (here, just the reference to falling rocks). A photograph accompanies the sketch (Figure 3.3). Vulnerability appears clearly in this photograph, with the road, and,

4 http://www.brgm.fr/sites/default/brgm/edutheque/kits_pedagogiques/reunion/fichier/fiche/Fiches_Risks.pdf, consulted January 27, 2016.

implicitly, the issues at stake, by the width of the road. The latter is assumed to be in line with the traffic and its importance as much economic and social as political. However, there is a major element missing here, and whose absence is all the more surprising considering that the BGMR is also a world of engineers: corrective works. They appear minimally in the photograph that the BGMR presents, with the protective wall. Illustrations of the definition bring to light the deficiencies, by lack of coherence. An essential component is absent from the definition of risk that BGMR offers, which is related to corrective works, and to what they achieve, at least in part.

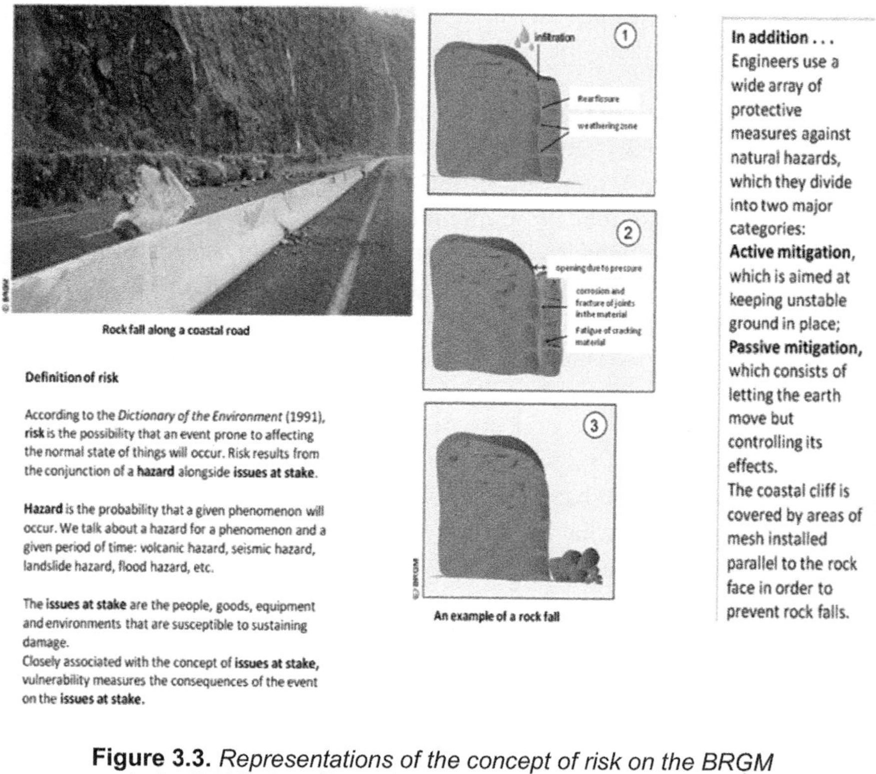

Figure 3.3. *Representations of the concept of risk on the BRGM website (2015). http://www.brgm.fr/sites/default/brgm/Reunion_kitpedago/fichier/fiche/Fiches_Risques.pdf*

Indeed, corrective works make visible a portion of the policies seeking to prevent damage and disasters. These works are sometimes even discreetly present on any simple panel displaying risk: here, we provide an example, located in Hong Kong (Figure 3.4). Moreover, during field studies, you would have to be very near-sighted not to see the corrective works today. We add that the land-use policies associated with disaster prevention are displayed and visible as well, while they are still very seldom included in the usual definitions of risks.

a) Photo taken facing uphill, Victoria Peak.

b) Photo taken facing downhill, Central.

Figure 3.4. *An example of the display of disaster prevention policies and their limits. Hong-Kong, Victoria Island, August 2011. The limits of the retaining walls are mentioned on the sign with the notice "Retaining wall pending upgrading works": the corrective/protective works are being improved. Photo: P. Pigeon*

This absence or quasi-absence reappears on the National Observatory of Natural Risks (in the original French: l'Observatoire national des risques naturels or ONRN) site, among others. Indeed, the site is linked to the desire to assess the policies that France has had in place with regard to disaster prevention since 1982. However, here again, the definition of risk that the ONRN provides addresses only the hazard/vulnerability binomial. Both the definition and its justification do not take prevention policies into account. They nevertheless make up the institution's *raison d'être*. We return to the inventory of the so-called "natural" hazards, which we assume makes it possible to define risks and disasters (Figure 3.5), as if the insurers were themselves censored by their own definition of risk.

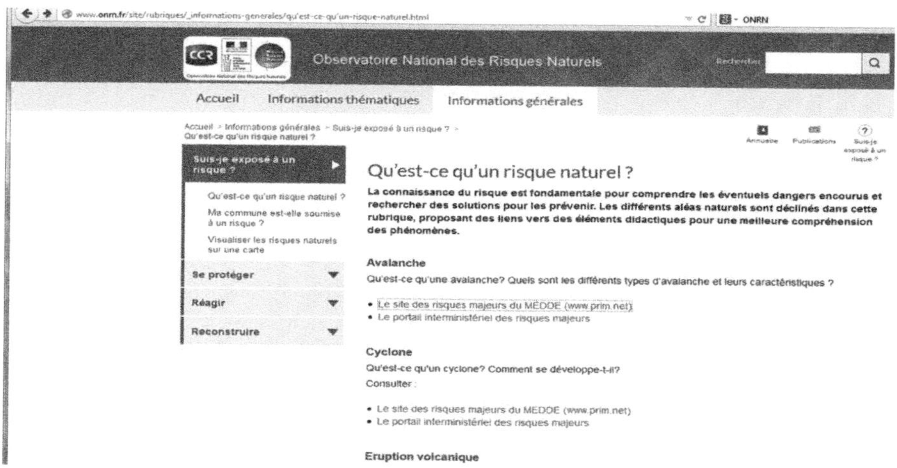

Figure 3.5. *The definition of risk according to ONRN and its declension on its official site (ONRN, 2015) Qu'est qu'un risque naturel? = What is a natural risk; Avalanche = avalanche; cyclone = cyclone; eruption volcanique = volcanic eruption*

However, part of the research recognizes this shortcoming. Certain European programs are beginning to integrate "coping capacities" into the basic definition of risks, as Figure 3.6 shows using an extract from Kuhlicke and Steinführer [KUH 10]). This is still not a common inclusion. For example, Shi and Kasperson [SHI 14, p. 4], in their monumental atlas on risks in China, establish that: "disaster losses and damages are consequences of the interactions of hazards (H), exposures (S) and the environment system (E) in which disasters occur". As we can see, the system's environment is too imprecise to be able to explicitly address the surprisingly small place that prevention policies occupy in the definition of disaster risk today. The very visible, and even more and more visible, remains almost invisible in the fundamental definitions of risk that we use today. This point was emphasized in Pigeon [PIG 91], and it remains relevant.

In view of the preceding, the conceptual models which we use today in the field of disaster risk reduction must also have numerous limits. These models are nevertheless the ones that are supposed to provide the most logical and coherent possible sense to the definitions and methods used in the dual field of understanding and preventing disasters. Deficiencies like these make it easier to understand why we lose much more when we have never tried so hard to prevent disasters. However, the principal question becomes even more obvious: why does this obvious lack, which has been identified, continue? Even though, as we have seen, disaster prevention policies contribute to effectively preventing certain disasters despite their limits.

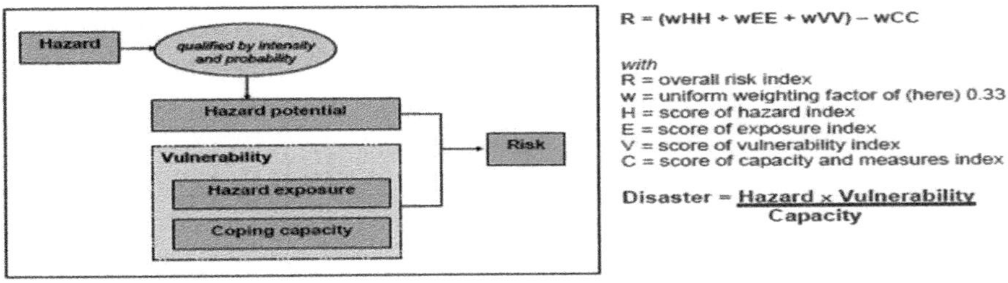

Sources: Greiving 2006 (left), Bollin and Hidajat 2006 (upper right), Davis 2004 (lower right)

Figure 3.6. *"Examples of graphical and numerical representations of (social) capacity in natural hazards research". A basic definition of risk in Kuhlicke and Steinfuhrer [KUH 10, p. 14]*

3.3. The difficult search for a conceptual model of disaster prevention

A conceptual model [CUT 08] or a conceptual framework [BIR 06] here refers to a graphical representation of the relationships that exist between the basic components of the definitions of risk that are, *a minima*, hazards and vulnerability or hazards and issues at stake. From this perspective, the model provides a total view, comprehensive in the etymological sense, of disaster risk. The approach is therefore fundamental, and, potentially, equally useful, since it should guide prevention policies.

However, since there is no agreement on the definition of risk, and even *a minima* [DER 09a], the challenge represented by conceptual models related to disaster prevention is colossal. It has already been identified by several authors such as Birkmann and Cutter. The latter, who are also engaged, in a revelatory way, in reflections on the management of databases on disasters [GAL 15], writes:

"Few researchers have attempted to combine all the factors that contribute to vulnerability, let alone measure them empirically. The most often cited conceptual models for hazard vulnerability include: (1) Blaikie and Wisner *et al.*'s pressure and release model [WIS 04]; (2) Turner *et al.*'s [TUR 03] vulnerability/sustainability framework; and (3) Cutter's hazards-of-place model of vulnerability". [CUT 96, CUT 00]

In their effort at summarizing, Birkmann and Cutter resume the attempts that are the most cited in the literature. They are possibly reproduced in the summary reports that international institutions publish. Here, we present some of these models and compare them with other more recent ones, by explaining why they all pose a problem. This way, we complete and continue the effort at analysis and reflection taken as much from databases as from definitions of disaster risk. On its own, the very rapid evolution of the models proposed shows their unsatisfactory character, such as the current impossibility of reaching a solution that is more useful and coherent at the same time.

Let us look at how and why.

The first model that we propose is taken from a reference work on the issue ([BIR 06, p. 23], Figure 3.7), and continues several previous attempts. He is very analytical, and systematically separates hazards, exposure, vulnerability and coping capacities. It is therefore not adapted to the very dynamic, and above all reflective, character of disaster prevention. It is moreover linear, following the arrow of time. It is clear that it cannot represent returns on experience, even if it seems apolitical. There is nothing after "disaster risk", as if future disasters did not even exist. Here, we might think that coping capacities actually incorporated prevention policies. However, these policies would then be outside of history, as if they were perfect. Where does the model integrate returns on experience? Here is a model of disaster risk, where the risk is outside of the long term, but indispensable to the concept of risk. The concept nevertheless imposes a projection in the future in hopes of reducing future damage.

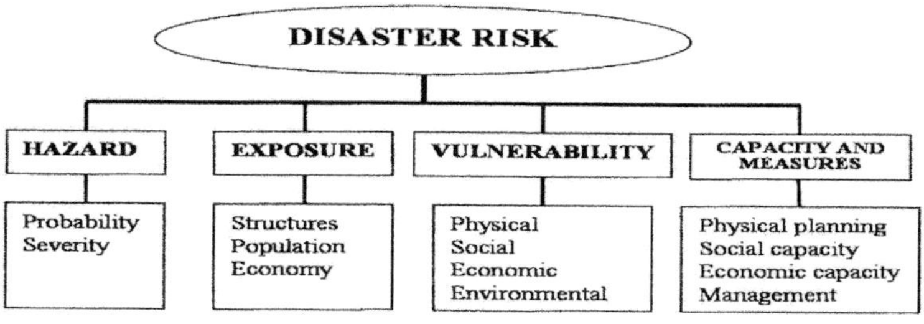

Figure 3.7. *The conceptual framework for identifying disaster risk [BIR06, p. 23]*

The second model, that of Blaikie *et al.* ([BLA 94], Figure 3.8), which is the first version of the pressure and release (PAR) model, integrates policies into the "progress of vulnerability". It addresses one of the weaknesses of the preceding models, since it explicitly shows the trend in human societies inadvertently contributing to disasters over time. In this model, the hazard has less of a place than the vulnerable elements. This model was used and discussed as part of the European program MICRODIS. However, we see that the hazards are still separated from the vulnerability, except at the moment of the damage/disaster, and that the returns on experience are not explicitly included here. This is the main reason that this model was revisited in the Wisner *et al.*'s [WIS 04] version (Figure 3.9). We will see that the representation is done, this time, in loops. These loops attempt to convey the returns on experience: the policies are revised after the damage/disasters. Policies are part of the history of disasters, they are even a major engine of this history, as much by their contribution to prevention as by their limits – and their conditioning, which we will come back to.

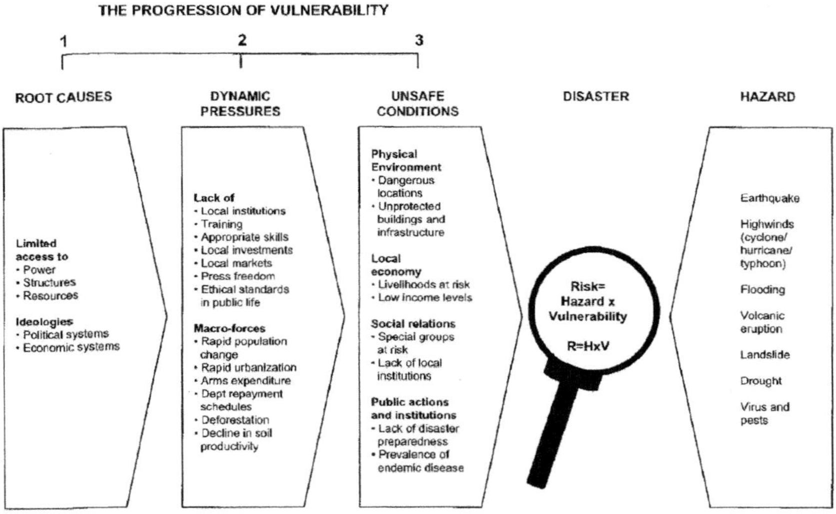

Figure 3.8. *PAR model: the progression of vulnerability [WIS 04, p. 51]*

However, there are still no relationships between hazards and vulnerable elements before the realization of the damage. This model is therefore not coherent enough with what can be observed during returns on experience or field studies. It does not consider corrective works, for example, which are also called protective works. Simply, a dike, a drain, a spur or an avalanche-prevention structure requires admitting that the hazard (watercourses, landslides, avalanches) is modified by the work. Managers hoping for a reduction in the frequency of events occurring as well as potential damage BEFORE the hazard and exposed and unequally vulnerable elements of the land occupation meet during the next event. The hazard has been modified before the damage, something that the model does not include. The models no longer take into account the fact that hazards, unequally vulnerable exposed elements, and prevention policies interact and co-evolve, at least in part. This is simply what the generations of corrective/protective works that can be identified during field studies (Figure 3.1) have revealed. Certain hazards do not only reveal the predisposition for damage during the occurrence of a flood, as shown in Figure 3.9. They are modified before the damage, and these modifications also contribute to explaining the damage. Is this visible in the model in Figure 3.9?

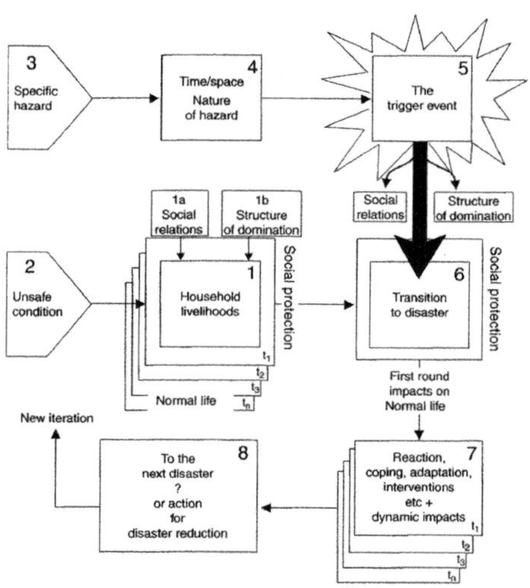

Figure 3.9. *The access model in outline [WIS 04, p. 89]*

The development of a portion of the research (especially on climate change) explains the evolution of the models toward the explicit inclusion of interactions between hazards and unequally vulnerable exposed elements. This is what we find in Turner's model (Figure 3.10). Just as for the models coming from schools of thought on socio-ecological systems, whether they belong to the group Resilience Alliance [GUN 02, ROC 14] or not [SHI 13, REN 13], their interest is in identifying and graphically representing feedback loops. The feedback loop makes it possible to integrate returns on experience into models, such as the limits of prevention policies. This mode of representation corresponds more to dynamic readings of disasters as well as the prevention policies that are associated with them. This time, these models finally integrate the existence of interactions between corrective/protective works, hazards and unequally vulnerable exposed elements. However, these models encounter numerous limits.

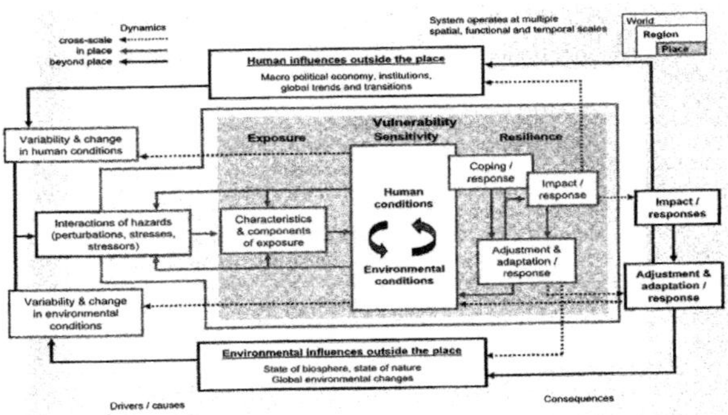

Figure 3.10. *Turner's conceptual model [BIR 06, p. 27]*

The price to be paid for the models associated with the interpreting disaster prevention policies as part of the socio-ecological systems is high in terms of readability. Moreover, it is currently not possible for the human capacity, direct or indirect, to affect "all" of the hazard, for example the deep-focused earthquakes. It therefore remains a part, although a limited one, of the factors contributing to explaining disasters which must remain separate from interactions between hazards and unequally vulnerable exposed elements. We should therefore continue to explicitly integrate natural hazards into models: these hazards influence human land occupations without the latter being able to influence them in return. There are hazards for which no form of human intervention, direct or indirect is demonstrable [PIG 05]. As a result, representation becomes difficult. These limits seem to represent, albeit wrongly, a return to the previous modes of understanding which completely dissociates "natural" hazards and exposed elements. If only the interlocutors were hostile to systemic formalization, they could easily claim that it was not so useful think so much to get back to "natural disasters", in the end. However, above all, it is about finding the most coherent possible solution for representation, and nothing more.

Finally, we can better understand why today we can find models such as those of Shi and Kasperson [SHI 14, Figure 3.11]. We can still be surprised to see a "disaster as system" represented without any explicit formalization of the relationships between major elements, here, *a minima*, the hazards and unequally vulnerable exposed elements, that form the basis of the definitions of risk and also of a system. These are nevertheless interactions, and even more, co-evolutions, even if they are partial, that, as we will see later, justify an interpretation in terms of the system.

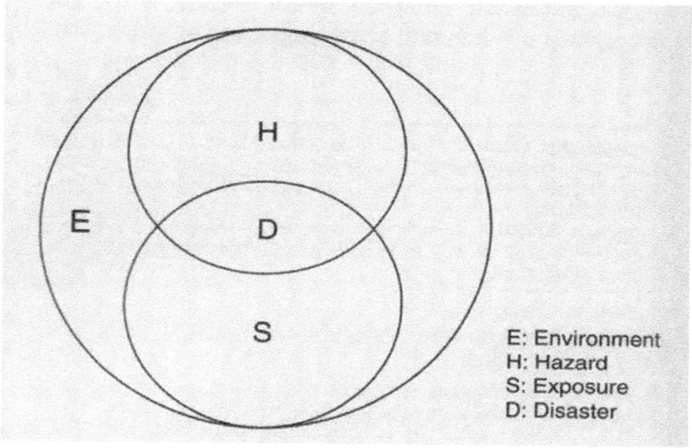

Figure 3.11. *Shi and Kasperson's conceptual model [SHI 14, p. 4]*

The extreme difficulty of finding a solution that can represent, in a readable way, the definitions of disaster risk, themselves very heterogeneous and controversial, can therefore be easily seen, while aiming for the most logical coherence possible. The fact remains that we find a poor relationship everywhere, presented marginally at best: disaster prevention policies. It is therefore useful to try to understand why this discretion is so evident.

It is equally tempting to look behind these desperate efforts to find a solution that is simultaneously clear, coherent and usable for defining disaster risk as one of the major reasons for the paradox that forms the base of this book. The limits of policies show the inability to recognize and identify certain key elements in the definitions, and especially themselves.

They can also show great reluctance when faced with modes of reflection moving explicitly toward forms of complexity.

Moreover, the basic definitions of risk are usually presented in the form of a product between two major elements. Simplification generally takes the form of a reasoning that assumes "all other things being equal", and it clearly aims at quantifying in an easy and useful way. This approach is currently accepted by physicists and chemists: "the perfect gas doesn't exist, it's a theoretical model that supposes that the molecules are not interacting with each other. It is not this way in reality. However, in numerous cases, the behavior of real gases deviates relatively little from that of a perfect gas. For this reason, the perfect gas laws are often applied to real cases" [ARN 98, p. 179]. On the other hand, in the field of research concerning disaster prevention, we have no difficulty in showing how observable situations deviate considerably from the conceptual models, calling them into question.

Can we get closer to complexity while remaining readable and operative in every case in the field of disaster prevention and reduction of disaster risk?

Before returning to this question in the third part of the book, we can only acknowledge the equally precarious nature of prevention policies in our own field, that of action and management.

4

The Limits of Disaster Prevention: Returns on Management Experiences

Researchers are confronted with problems posed by the definitions and models that they use to give as much coherence and usefulness as possible to the prevention of disaster risk. However, for their part, the managers and the political decision-makers, who use the same definitions, eventually recognize the limits of actions decided in order to anticipate future damages and to reduce their intensity as much as possible. The emphasis put on limits of management equally justifies the interrogation that underlies this book. It can also receive feedback in the research and reflections on political actions. According to Myriam Revault d'Allonnes [MYR 02, p. 35], "policy is created inasmuch as it reflects, where people make policy thinking about (and because they think) what they are doing". The philosopher concludes that "an essential function of political activity (…) consists exactly of confronting fragility, living with it, and not by controlling it by exceeding or abolishing it" [MYR 02, p. 154]. Indeed, multiple post-disaster feedbacks, such as reflections on reconstruction, confirm on their own the impossibility of finding a fully satisfying solution (and for whom?) to the problems that disaster prevention poses. These interrogations of the limits of prevention policies are exposed at the worldwide scale (section 4.1). We follow this with two studies that are more specific: one about the identification of dike risk (section 4.2), and the other on the analysis of prevention policies carried out in Ecuador (section 4.3). These will show that the problems that the disaster prevention policies pose are clearly identifiable whatever the differences between the land occupations implied, as well as the scale of the analysis or the institutions and even the local actors who are involved.

4.1. Returns on experience demonstrate the universal character of the limits

The very idea of return on experience reflects the interest in evaluating the differences between what was expected from a policy and what is observed *a posteriori*, that is after a disaster has occurred. The return on experience therefore highlights the limits of policies dedicated to disaster prevention, as well as their more implicit contribution to prevention. Return on experience especially makes sense when the discrepancies between the expectation and what is observed are large. In the eyes of the political decision-makers and managers, it is about understanding why, and seeing how we can hope to reduce the differentials of future events by returning to the measures taken. The multiplication of returns on experience is therefore involved in the pessimistic interpretation of disaster prevention policies, because it encourages considering the limits rather than the contributions of these policies.

In addition, the multiplication of returns on experience also encourages, necessarily, drawing attention to the limits of the capacity to learn. This kind of interrogation is very present in the reflections on the transition between the Hyogo and Sendai international frameworks on disaster prevention. This transition is justified by the limits that disaster prevention policies always encounter despite the prevention efforts encouraged and accompanied at the global level by UNISDR. The second priority measure that the Sendai Framework for Action mentions is dedicated explicitly to "strengthening disaster risk governance to manage disaster" [UNI 15a, p. 17]. The same approach justifies European research programs such as CAP-HAZ net or KNOW4DRR. These programs associate as closely as possible the managers of policies dedicated to disaster prevention. We can find the same interrogation with the International Forum of Antalya, which lead to the eponymous declaration in March 2015. This forum gave a prominent place to designers and managers of prevention policies, for example, those who work for the Red Cross, the Red Crescent or as representatives of NGOs. The declaration seeks to identify what needs to be met to effectively learn more from the aftermath of disasters, by positing that these lessons are not finally taken into account despite the returns on experience. This group of messages is clearly present in the GAR Report [UNI 15b] with entries on the "limits of control" [UNI 15b, p. 33], "unfinished business" [UNI 15b, p. 39], "reducing disaster losses, a partial success" [UNI 15b, p. 43].

A particular element can show the limits of disaster prevention policies, mentioned in returns on experience: that of reconstruction. The GAR Report [UNI 15a] devotes no less than 8 pages to it, introducing it with a revealing title: "Business as usual or building back better?". The title was officially translated into French as: "maintenir les mauvaises habitudes ou reconstruire en mieux?" The introductory heading of section 8.3 of the report, which is dedicated to this "desire to build back better" is very clear: "However, in many situations, the desire to build back better quickly gives way to business as usual". We can draw attention to the way the theme is presented differently between English and French versions. The English version points out more explicitly that the inclusion of disaster prevention during reconstruction calls into question the standard ways of life. "Business as usual" could be more faithfully translated into French as: "train-train quotidien". Particularly strong motivations are therefore needed to consider a more structural challenge to "business as usual", and the related land occupations, which existed before the disaster.

The limits of the desired adaptation once again draw attention to the importance of the major causes, essential ones, of obstruction. It involves *root causes* that a part of the Anglophone literature on the disaster risk reduction mentions. Disaster prevention policies do not sufficiently take them into account during the post-disaster reconstruction. We can present this deficiency in the form of the following dilemma: despite the experience of the disaster, failure to reconsider the structure of the land occupation that existed before the disaster could mean even more and greater damage. There could also be more damage (nevertheless, what damage and for whom?) in the case where the factors of socio-economic inequalities were not considered, just like the power relationships that are associated with the structures of land occupation that existed before the disaster.

Worse still, emphasizing the limits of research and prevention policies can encourage even less calling into question of the structures of the land occupations that existed before the disaster, and that essentially contribute to causing them. In an uncertain situation, it may be more useful and more reassuring to return as much as possible to the past, for households as for many institutional actors, whatever the scale of their interventions. Let's not forget that poverty considerably reduces one's options, and provides encouragement to work toward a previous known state, that is to return to it.

What is the meaning of risk, and even more of uncertainty, for the embankment dwellers of Delhi who survive on the banks of the Yamuna, or for the inhabitants living precariously in the informal settlement of Goma in the Democratic Republic of the Congo? Following the results of the survey conducted by Trefon [TRE 13] and presented during the Gisenyi Colloquium in 2013, as part of the GeoRisCa/AVCOR program, the significance of the concept of risk is very limited for these populations of refugees living in extreme insecurity. Some of the people surveyed living in the informal neighborhoods of Goma even stated that they did not have a future.

Therefore, the question involves finding the best possible way to integrate the prevention of a specific type of damage, among a range of other possible damages, into these settlements during their reconstruction after a disaster. This also involves knowing what can be done in terms of prevention when the fundamental conditions of settlements are especially unfavorable. This is particularly the case when the factors of uncertainty are very high, or even in case of the significant presence of very poor populations. Rebuilding settlements involves structuring these according to other types of damages, or interests, despite the disaster that has occurred, and according to the type of damage which is considered more important, at least by a portion of the social actors.

The expression "identical reconstruction", which is discussed in particular by Moatty [MOA 15], corresponds to the desire to forget the disaster experienced and to erase its effects. This form of catharsis is addressed by psychology as cognitive dissonance. According to Festinger [FES 57], cognitive dissonance explains the decision, the behavior and the action that are conducted as if a type of risk does not exist while the damage of the disaster has been lived by the people involved. These people nevertheless seek to deny this. The literature is full of these examples, as is the history of disasters, especially in terms of urban disasters. On this point, we recommend reading the works of Alain Musset [MUS 02] in *Nomadic Towns of the New World*. Despite the awareness of disasters associated with earthquakes and floods in Mexico, the successor to the Aztec capital Tenochtitlan is still located on the same site. Political projects to relocate the capital were, and still are, met with resistance from the populations and institutions that oppose it despite the awareness of disasters. It is common to find similar reactions during field research, when a person surveyed, who has lived through a disaster, does not hesitate to announce that, for her, moving or even having to adapt their property constitutes even higher, and unacceptable, damage. One of the authors of this book also heard a representative of a citizen's defense association tell French deputy Guy Geoffroy, during a working session of the Steering Committee for the Prevention of Major Natural Risks (in the original French: Conseil d'Orientation pour la Prévention des Risques Naturels Majeurs or COPRNM) at the National Assembly in May 2010, that risk, for her and the citizens that she represented, was not directly linked to the River Oise's flooding. According to this interpretation, it is actually the Risk Prevention Plan (in the original French: Plan de prévention des risques or PPR) with its zones that are legally enforceable and hinder construction that constitutes the risk. Without explicitly addressing this remark, as provocative as it is revealing, Deputy Guy Geoffroy clearly mentioned in the report on the session: "I do not want to take a position in the name of the working group but I would like to remind you that the PPR is a document created by the State and not requested by the inhabitants" [COP 10]. The cognitive dissonance illuminates the very unequal acceptance of disaster prevention policies. The decision-making process and the behaviors that it describes favor the reconstruction of land occupations after disasters as identically as possible. However, these decision-making processes cannot only involve individual choices. This would suggest that the discrepancy observed between disaster prevention policies and returns on experience could be reduced by a correction of supposed individual errors of assessment, by a simple process of education.

Nevertheless, the returns on experience are effectively included in disaster prevention policies. According to Moatty [MOA 15, p. 339], "'good reconstruction' is a question of perception, point of view and ranking of territorial issues". That is what the thematic of dike risk can show. Based on the multiple post-disaster returns on experience in France, dike risk reconsiders these risk prevention plans, which are management tools associated with Tazieff's law (from 1982). However, as we discuss in the following, the inclusion of dike risk in planning documents does not allow more elimination of any limits to this policy, despite a critical revision.

4.2. Dike risk in France: an example of unintended consequences

A dike is a civil engineering work that has the official goal of correcting a watercourse or avoiding inland or maritime floods, and therefore to protect land, buildings, activities, populations that are found nearby. These works are used in numerous countries, where they are even considered vital. That is the case in the Netherlands, with its Delta Works [GAR 15, WAR 13]. They are still advocated by development aid programs, such as in Bolivia, Indonesia or Bangladesh. It can therefore appear surprising, and even illogical, that a protective work that is supposed to reduce risk, and which is disseminated and advocated worldwide, becomes a factor in increasing the same risk. Yet it is nevertheless this reversal of interpretation and practice that is described by the expression "dike risk", as a result of multiple returns on experiences. The expression explicitly acknowledges that the disaster prevention policies that rely on these corrective and protective measures contain within themselves their own limits. However, what is dike risk specifically?

In France, "dike risk" was officially recognized by a circular of 2002[1]. The national flood risk management strategy developed by the Department of Risk Prevention [DGP 14] mentions that "keeping the populations protected by existing works requires long-term maintenance of these according to the objectives pursued by the national plan for rapid flooding". It was nicknamed the "dikes plan" and concerned both inland and maritime dikes. Dating back to February 17, 2011, it "allowed the development of a major initiative relating to the reliability of the works and the systems of protection" [DGP 14, p. 13]. Legal developments therefore encourage the official recognition of dike risk. In fact, it urges reconsideration of the relationships between corrective works and regulatory zones that display the level of risk on risk prevention plans, since Tazieff's law [TAZ 82] and its enforcement decree of 1984. A corrective work could then, traditionally, justify the classification as a blue zone, with average risk, and constructible, while, without a corrective work, the same zone would have been classified as red, which means inconstructible. This is no longer necessarily the case today.

1 The circular of April 30, 2002, relating to the State policy with regard to foreseeable natural risks and management of spaces located behind dikes protecting against floods and marine flooding http://www.bulletin-officiel.developpement-durable.gouv.fr/fiches/BO200219/ A019 0045. htm, consulted January 14, 2016.

This is what has been observed. For example, the prefecture of Gironde [PIR 14] officially integrates dike risk into the maps showing the risk areas of a Risk Prevention Plan: “the revision also makes it possible to integrate the risk of the dike breaking, which, until then, had not been included: the breaking of a dike caused by extremely high water generates particularly dangerous currents directly behind the work. This ‘over-hazard’ is now included in the new PPRI, making the zones in proximity to dikes inconstructible”. Specific red zones that prohibit construction on the land concerned convey dike risk concretely. These zones often take the form of strips marking the layout of the dikes on the maps.

The prohibition of building on these sites explains why dike risk is strongly contested by a number of citizens, especially the owners of the land directly concerned, and also of local governments. Indeed, these citizens and mayors feel that the dikes are not only necessary, but above all that they are sufficient, and, in addition, very useful. They contribute to making it possible to build on sites that were previously flooded too frequently to allow development without particular precautions. Urbanizing without dikes requires additional costs and multiple inconveniences. Dikes, just like all of the other works addressing hazards, provide the illusion of the disappearance of hazards, and the damage that is officially attributed to them. We better understand the success that, despite multiple denials, this type of traditional approach to risks has had and still has, and specifically the risks considered “natural”. This approach has sometimes been called hazard centered by geographers and sociologists [GUE 14a, GUE 14b].

In fact, these corrective works, which are also called protective works, transform the relationship between the frequency of events and the intensity of the damage. This is particularly the case if the buildings are constructed near these works and without taking into account the transformation of this type of risk. The risk is in no way eliminated, or even residual: it is transformed, and it can even increase over time. With this reading, which relies on very numerous returns on experience, corrective works contribute inadvertently to the formation of disasters. As a result, the opposite of the goal that the prevention policies assign to the works, both corrective and protective, is achieved.

It is therefore understandable that the French State, or other institutional actors in risk prevention, such as insurance companies, draw attention to dike risk, and seek to prevent it. Conversely, a number of local actors, and especially the owners of land near dikes or corrective works, highlight the role of dikes and of any other works also considered protective. Their attitude corresponds well to that which cognitive dissonance describes, especially when people have personally lived through a disaster. Put more simply, for them, there would be more damage in not continuing to believe in the absolute protection of these works, or in admitting the limits of policies aimed at preventing a type of disaster. One of the authors of this book was able to verify this during a conversation with the mayor of La Faute-sur-Mer, in June 2010, after the disaster of February 2010. The mayor affirmed his absolute confidence in the dikes that, effectively, did not fail on the site where 29 people drowned in their homes. The photograph in Figure 4.1, taken at low tide, confirms the mayor's statements: the dike did not give way on this site.

Figure 4.1. *Aerial photograph of the area of La Faute-sur-Mer with the highest concentration of mortality, February 2010*

However, the same photograph also shows, unfortunately for the 29 people killed, the limits of the absolute confidence in protective works. Indeed, after February 2010, in spite of the gruesome disaster, the mayor spontaneously declared his absolute faith in dikes, just as some of the inhabitants did. The dike prevents the drainage of water trapped in the depression and also increases the time that the buildings are submerged in brackish water. Above all, the mayor seemed to forget that the presence of the dike had justified, in part, the construction of housing developments on sites that were officially acknowledged to be at risk for floods according to reports on deliberations of the City Council [PIG 12b]. Here, we find the necessity of not limiting analysis to the individual decisions only (for example, living behind a dike), which is partially accounted for by the theory of cognitive dissonance. The disaster of February 2010 conveys collective choices, according to the different lobbies that were moreover represented in the City Council. Disaster is fundamentally prepared for by power relationships over time, and it is above all political.

The reconsideration of disaster prevention policies does not result, therefore, in eliminating the limits. The price to pay for integrating dike risk into risk prevention plans, in cases of very vocal opposition by a portion of the local actors directly concerned, is known. It shows the local opposition to the application of policies considered top-down, here, taken in Paris and/or in the prefecture, Jacobin measures coming from the capital. In the Vendée, traces of the Chouannerie (a royalist uprising of the late 18th Century), can reappear (Figure 4.2). Moreover, this opposition is often justified by the contestation of expertise on the basis of the limits of available knowledge. Uncertainty is therefore used as a reason for opposition. Opposition to disaster prevention policies, even when reconsidered in light of disasters experienced, is also fed by the local capacities for response, which can be limited. The question is posed for a commune such as that of La Faute-sur-Mer. Moving, even with compensation, was considered unacceptable by several people surveyed. Finding a similar quality of life, near a recreational harbor, which appears in Figure 4.2, would require higher property and tax costs that would be hard for some retirees with modest incomes to afford. One person even compared La Faute-sur-Mer to the communes on the île de Ré, where unfortunate retirees are being pushed out, in particular due to financial pressure connected to the cost of the property.

Figure 4.2. *Local opposition to disaster prevention policies associated with prefectural and Parisian authority. Photograph taken in La Faute-sur-Mer, June 2010 (Photo P. Pigeon) Translation: 'Mr. Prefet, let us make our choice. It's our freedom'*

These different elements, among others, explain why we are currently seeing numerous significant discrepancies between the requirements for corrective works and/or architectural adaptations that the regulations in risk prevention plans contain and their effective implementation. These discrepancies even concern areas for which the poverty of the households is not a major characteristic, such as in the lower part of the avalanche channel of Brévent in Chamonix. Labeled a blue zone of the PPR, this part of the commune of Chamonix highlights a discrepancy between the architectural adaptations required by the PPR and what is actually carried out: "close to 75% of the buildings in the hamlet of Brévent do not seem to follow the regulations of the Chamonix Avalanche Risk Prevention Plan" [CAL 12, p. 122]. It is recognized that the architectural requirements have the goal of reducing the intensity of future damage, such as for example, in this case, the reinforced and blind building walls, without windows, that are facing uphill and exposed to avalanches. Opposition to the prevention policy by contesting the zones officially displaying the risks contributes to explaining why such a low percentage of owners, who are involved, consider the architectural requirements. Conflicts concerning zoning risk prevention plans and their regulations counteract disaster prevention, and also demonstrate the limits policies meet.

Fundamentally, dike risk, which legally recognizes the limits of corrective works, tends to activate opposition to disaster prevention policies because it challenges previous political choices. The latter were more accepted by the owners of land next to the dikes, such as by the various lobbies associated with the construction sector. However, they favor the inadvertent contribution to disasters over time, as the tragic and iconic example of La Faute-sur-Mer in February 2010 demonstrates.

This is not a French exception, even if the dike risk is legally integrated into a national policy whose equivalent we have not found worldwide. Indeed, recognizing the limits of corrective works exists elsewhere, in Bolivia in particular [HAR 13], or in Indonesia [MOA 15]. It takes other forms, for example, in Ecuador, and in connection with disaster prevention associated with eruptions of the volcano Cotopaxi, in the strengthening of civil safety policies [ROB 09]. Moreover, the limits of dikes as a means for preventing floods have been known for a long time ([BUR 01, KAT 86] and even [WHI 45]). Just as fundamentally, these developments draw attention to the limits of sectoral policies, those that are based on a single element in hopes of reducing, and even, for utopians, eliminating future damage. Yet, disaster prevention policies cannot focus on a single risk factor, or on a single managing institution, or else they will find their limits even more quickly. It is moreover the main reason why segmentation of prevention policies by several institutions involved poses so many problems, which are also recognized.

4.3. Segmentation and politicization of risk management in Ecuador

The case of Ecuador underlines the weight of a larger context in the understanding of disaster prevention policies. The latter encompasses diverse problems (social, political, economic or institutional) which can *a priori* seem foreign to them, to which they can never be closely linked. Moreover, within the national risk management strategy currently developed in the country, several visions of risk policies coexist. This confusion urges us to consider disaster prevention not as a series of transposable technical measures, indifferent to context, but as policies which must be contextualized [REB 12].

4.3.1. *Emergence and ambiguities of a national strategy for risk management*

In 2008, Ecuador had just gone through a long period of social unrest and political instability that had begun in the middle of the 1990s, and was at the threshold of significant institutional, economic and political changes [RAM 10]. Following this troublesome period, a new constitution was adopted. It made risk management a mandate of the State (art. 389) as well as a prerogative of the local public authorities (art. 390) as part of the decentralization process that the country was undergoing. Concurrently with this regulatory process, there were heavy rains pounding Ecuador's coast over several months. The crisis situation speeded up the creation of a Technical Secretariat for Risk Management in April 2008. The Secretariat took over the prerogatives of the two institutions in charge of civil protection (the Civil Defense and National Security Council of Ecuador), and was responsible for setting up the Decentralized National Risk Management System (SNDGR) set out in the Constitution.

The public authorities (from municipal to national) must include disaster prevention measures in their management, and the technical secretariat is the guarantor. The national government's initial will defends the transversality of risk management, the respect for the subsidiarity principle, and its urgent nature in the context of a coastal crisis in 2008. A regulatory disposition of January 2014 even provided for the removal of mayors in case of negligence in the distribution of land use authorization that does not take risk conditions into account. However, in practice, the setup of this national strategy proves not to be as consistent as it is advertised to be.

The technical secretariat in charge of the SNDGR became the National Secretariat of Risk Management in September 2009 (with ministry status and representation in each one of the 24 provinces in the country), then the Secretariat of Risk Management (SGR) in August 2013. The change in the Secretariat's institutional status weakens its actions dealing with public and private organizations in the country, and we can even read the last reform as an institutional demotion, with representation in only nine of the country's planning area. Moreover, to date, the country does not have a specific law on risk management. For the national experts consulted, the dispersion of tools and measures currently present in Ecuador's legislative arsenal does not help to carry out a consistent risk management policy at different scales. Despite this, the Legal Council of the Presidency continues to oppose a "law of risks". However, it is true that a new law calls for new skills, and as a result the transfer of additional funds as part of the current process of decentralization [REB 15].

4.3.2. *The multiple obstacles to setting up SNDGR*

Even if, within Ecuador, the SNDGR initiative is extremely valuable [OCH 15], its difficult and unequal application over the entire national territory encountered obstacles and resistance at the political, institutional and human levels. We will present certain examples, related to the weakness of public institutions still in the process of consolidation, the continuing weakness of the skills of the civil servants in charge of questions of risk, the politicization of the setup of risk policies, or even tension, for the SGR, between local application and national coordination.

The adoption of SNDGR happened at the same time as the decentralization process that began in 2008. Ecuador's national planning agency (SENPLADES) reported that, in 2013, among the assessment of 41 out of 221 of the country's local governments, a third did not have a planning or land management and planning service. In addition, a third of the municipal development plans available in this sample had no operational provisions [SEN 13]. The weakness of the capacity of municipal institutions in the area of planning has consequences for the integration of risk management into land management, which happens often through regulatory control of land use.

Moreover, the decentralization in progress and the new duties taken on by the local municipal powers, responsible for the setup of risk management in their jurisdiction, suggest difficulties in terms of human resources. People (civil servants, elected officials or technicians) who are responsible for risk management locally, when the service exists, show a very heterogeneous set of skills in the field. The SGR even provided campaigns for informing agents of the existence of budget lines to which they can claim as part of setting up local risk management.

In addition to these institutional and human obstacles, there is the political rivalry between the national government (and the partisans of President Rafael Correa) and the local majorities held by the opposition. The most populous city, Guayaquil, is directed by a proclaimed opponent, Jaime

Nebot, which accentuates the rivalry between the two largest cities in the country, with Quito the seat of national power. The political animosity is doubtlessly not foreign to the attitude of the SGR, a national institution whose head is named by the Presidency of the Republic, and then in 2013, the latter boycotted a joint crisis simulation exercise between the municipality of Guayaquil and the Office of US Foreign Disaster Assistance (OFDA). Moreover, it is important to indicate that since September 30, 2014, the cooperation agency USAID, which the OFDA depends on, is no longer authorized to cooperate with Ecuador in the field of risk management, after 25 years of activity.

The multiplicity of municipal risk management initiatives is difficult to coordinate. Especially pronounced between Quito, Guayaquil and Cuenca, this heterogeneity is found in different degrees throughout the entire region and tends to show the breaking up of initiatives. Beyond the SNDGR, these are the limits (along with, of course, opportunities) that the decentralization process presents, which are reflected in the difficulties setting up the national strategy for risk management. The common challenge therefore remains national coordination, in tension with the principle of subsidiarity and the autonomy of local governments.

4.3.3. *Different visions of risk management coexist*

However, in addition to these different obstacles and the territorial fragmentation of risk management initiatives, several visions of risks (and their management) coexist within the SGR, and within the country. These visions reveal profound differences in the conception of risk, causality, and the types of policy to be developed.

According to a civil servant working for the Directorate-General for European Civil Protection and Humanitarian Aid Operations (ECHO) and posted in Ecuador in 2014, the actors in international cooperation on risks now position themselves with regard to an underlying trend that consists of going from hardware (protective works, delivery of equipment) to software (training, transmission of expertise, co-construction of policies). We saw in Chapter 2 how the impact of engineering solutions, along with international cooperation, remains important. However, the trend of orienting cooperation initiatives toward software hides another more general dividing line that concerns the content and objectives of training and coaching, and especially the political scope of risk management.

Many cooperative actions involving software fall within the category of preparation activities for crisis management. The creation of community brigades is promoted, in order to multiply "empowerment" workshops, to train and sensitize communities to favor "appropriate behavior" in different local rural governments, as is the case as part of the FOCAPRE program financed between 2011 and 2013 by the OFDA using close to a million dollars, and run locally by the NGO CARE. Furthermore, other operations, in particular those supported by the PNUD-Ecuador, emphasize the importance of the institutions becoming involved in entire sectors of local public action along with the efforts regarding individual capacities or collective training in crisis preparation. For this latter vision of risks, of their creation and their management, it is reinforcement of the capacity of institutions to consolidate regional planning and favor prevention that is a priority. The DIPECHO programs (whose financing depends on ECHO) coordinated by the PNUD in 2011–2012 and 2013–2014 are headed in particular in this direction [REB 16].

We therefore find two different conceptions of risk management and of initiatives to carry out: on the one side, the priority given to response and preparation, and on the other side, the attention centered on the capacity for development and planning. These conceptions are not necessarily incompatible, but they demonstrate an unresolved debate within the national risk management strategy in Ecuador. Should we anticipate and prepare for a crisis as much as possible by setting aside budgetary reserves to "face" disasters, or should we devote this money to the problems of development? The different options respond to choices and political positions that are seldom discussed. Yet, risk policies are not simple technical arrangements. They correspond equally to values that it is important to understand inasmuch as, in management practice, they make up the limits (not technological but political or axiological) to adoption of policies, or recurring patterns of incomprehension.

The limits of disaster prevention experiences can be cognitive and conceptual, or even institutional and political. But, more than just the obstacles, they indicate the human possibility of modifying or shifting disasters, while admittedly not being able to eliminate all of them entirely, and even sometimes inadvertently contributing to causing them. A large part of the problems related to disaster prevention policies come less from their multiple limits (and difficult to overcome) than from the segmented and technical understanding of disasters, centered on the latter.

The problem that the limits of disaster prevention policies pose therefore requires decentering, which moreover began with the systemic approach. The issue therefore consists of creating courses of action to respond to the two questions that shape this work: why is it utopian to be able to eliminate every disaster? What can we do to manage and think about disaster prevention as best we can?

PART 3

Why and How Does the Prevention of Disasters Necessitate Thinking and Acting in an Imperfect World?

"*Difficulties [...]arise from our professing to reconstruct reality, which is the tendency and consequently mobility, with percepts and concepts whose function is to make it stationary.*"

(Introduction à la Métaphysique[1]*, [BER 34a, p. 111])*

The difficulty of pushing the limits that disaster prevention policies encounter has been considered on numerous occasions, as much by decision-makers as by researchers. Here, and in light of the previous two sections, it takes the form of the following question: how much more can we prevent disasters than we already do, without ever being able to completely eliminate them? This question originates from the following paradox, the examination of which constitutes the basis of this book: the more we know about it, the more we can manage, and, taking the argument further, the more we manage, the more we lose. This paradox justifies multiple research projects, as we have seen in the previous sections. The main response to both this paradox and the associated questions is taken up in the texts, which justify the transition from the Hyogo Framework for Action to the Sendai Framework for Disaster Risk Reduction. These are global policy frameworks which aim to accompany global scale disaster prevention policies. They rely upon research summaries, which dominate the field of disaster risk prevention (UNISDR – the United Nations Office for Disaster Risk

1 This translates as "Introduction to Metaphysics".

Reduction [UNI 15b]). The following response is provided: without having the ability or the desire to reconsider in-depth the social and political structures of human societies and their respective evolutions, it will not be possible to prevent disasters any further than we can at present. This approach has already been defended for a long time by the so-called radical interpretation of the issue, indeed that of the social construction of disaster risks [BLA 94, GAR 05]. According to this approach, the key to the issue lies within socio-economic inequalities, as well as within unequal power relations that persist within human societies.

Moreover, this manner of apprehending the issues posed by disaster prevention partly supports certain views upon the links between justice and inequalities, such as those of Sen [SEN 09, p. 13]. He clearly states that "*...a theory of Justice that can serve as the basis of practical reason must include ways of judging how to reduce injustice and advance justice*". He lays down the possibility of reducing injustice. However, Sen explicitly mentions on page 143 the "inescapable plurality of competing principles" during decision making. Reducing injustice does not only direct such decision making. Even more important, the search for justice may "*...lead to an impossible result, yielding an impasse*". It seems to us that we then uncover the basis of the issue: that disaster prevention policies are necessary. However, faced with numerous factors and the multiplicity of types of damage and interests to manage, it is just as necessary to acknowledge the limits of temporarily adopted policies. It is a matter of seeking to push back these limits, without being able to entirely eliminate them. Moreover, where does this "inescapable plurality of competing principles", which is indeed a barrier to reducing injustice, fundamentally come from according to Sen?

The so-called radical approach presents fundamental elements to understand the limitations of disaster prevention. However, even if this approach has been largely debated and discussed by existing research, it does not appear to be in a position to explain why we find so many discrepancies between risk definitions, available conceptual models, and arguments, indeed however simple, that the interpretation of multiple case studies arouses. Then it seems to us both useful and legitimate to also turn toward an epistemological register, and thus toward the limits associated with the ways issues linked to disaster risk prevention are established and thought through. The gamble may perhaps appear daring, and even conceited, given that we are not philosophers. We become aware of the increased risk of misinterpretation and also of anachronism, as soon as we borrow a quotation from an author such as Descartes or even Bergson claiming that it may clarify the current issues which preoccupy us. However, do not the limits of disaster prevention policies find themselves within those which are imposed upon humanity as soon as it endeavors more generally to control and to think? Raising this challenge forces us to turn towards both epistemology and philosophy. Moreover, this forces us to defend the possible contribution of philosophy to the understanding of problems and limits that disaster prevention policies specifically create.

Consequently, we now set out in this Part an interpretation that contributes to explaining both the necessary, and necessarily limited, character which frames all disaster prevention policies. The structure of this next chapter takes three stages: that of the necessary analytical dissection of issues (section 5.1), that, which is indeed complementary, of the systemic reconstitution of issues (section 5.2), and that of the limits which all analyses encounter, whatever form such analysis takes (section 5.3). The main question seems entirely transformed to us, as it imposes the need to find the means to prevent disasters while maintaining acute awareness of the impassable character of given limits. That, returns to the necessity of controlling and thinking within the confines of imperfection, while the extent of such imperfection cannot be totally reduced. It is thus by this acknowledgment that we turn it into an asset for action: reducing the limits without being in a position to eliminate them. Next comes the question which is developed within the last chapter of this work (Chapter 8), which is what should we do for the best?

5

The Principle of Cognitive Limits: Its Application to Disaster Prevention

5.1. The analytical basis: the Cartesian method and the fragmentation of notions

Initial light may be shed by the Discourse on Method[1] [DES 37]. It establishes several principles, some of which allow us to justify why defining the notion of risk, even today, involves fragmentation into several sub-notions. For risks, it is mainly a question of hazard, exposition and vulnerability. The approach is thereby systematic.

It may appear surprising to resort to a line of thinking which is already extremely dated, with the aim of better understanding current issues which arise from disaster prevention policies. A justification may be provided by a lexical and proxemics-based analysis of the term "systematic". On its website, the CNRTL (the *Centre National de Ressources Textuelles et Lexicales,* French National Center for Textual and Lexical Resources) places the term "Cartesian" within the group of prevailing synonyms (Figure 5.1).

1 The full title of this work is "Discourse on the Method of Rightly Conducting One's Reason and of Seeking Truth in the Sciences".

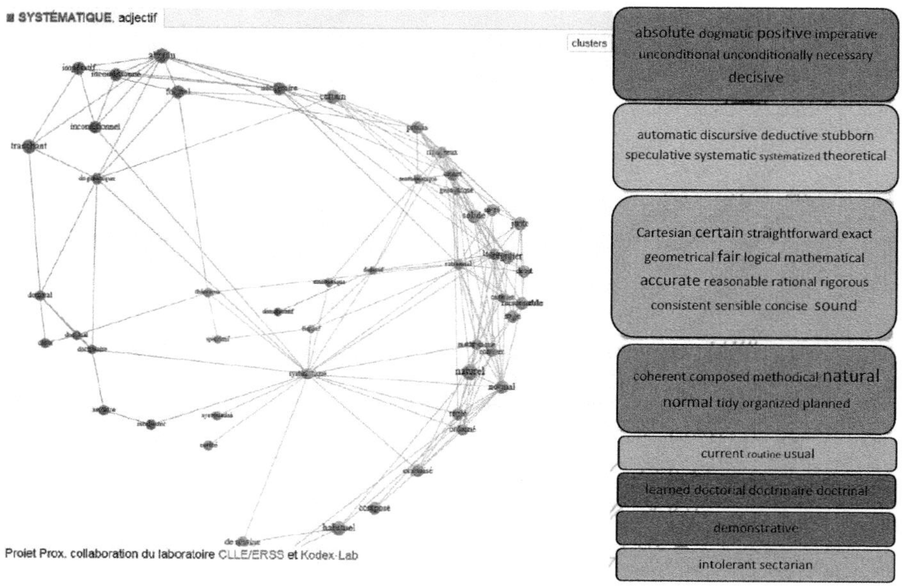

Figure 5.1. *Lexical and proxemics analysis of systematic terminology by the CNRTL (Centre National de Ressources Textuelles et Lexicales)*

COMMENT.– *Proxemics analysis extends to linguistics, the study of positions relating to interlocutors: in the first instance, proxemics appears interested in spatial relations between given subjects, in this case the lexical forms between them. This justifies the statistical studies of synonymy networks.*

Figure 5.1 shows the results of the proxemics analysis groupings (clusters) according to the so-called distance between the various synonyms and the term systematic.

Indeed, with the second principle of the method, Descartes sets out the necessity "...of dividing each of the difficulties under examination into as many parts as possible, and as might be necessary for its adequate solution" [DES 37, p. 46].

If we accept that the "difficulty", in this case, arises from the definition of risks, resorting to "fragments" of "difficulty" such as hazard, exposition and vulnerability, it may then be meaningful. We also note that Descartes refers to the limits of the approach, which the second principle advocates. It is not a question of actually resolving the "difficulties" but simply of "better resolving them". The requirement here is to find the least limited of all the potential approaches.

The second principle appears to initiate the third, despite the change in vocabulary used by Descartes to express his thoughts. He states that it is necessary to "start" from the simple and indeed from the "easiest to understand", and to "rise up" towards "composite" thinking. In fact, the third principle states the necessity "to conduct [his] thoughts in such order that, by commencing with objects the simplest and easiest to know, [Descartes] might ascend little by little, and, as it were, step by step, to the knowledge of the more complex; assigning a certain order even to those objects which in their own nature do not stand in a relation of antecedence and sequence" [DES 37, p. 46].

Again, it appears possible to defend, without misrepresenting Descartes's thinking while applying to our specific issues, that the "easiest to know", at least upon initial interpretation, is the hazard. It moreover comes first within the segmentations, which serve to define the notion of risk. In a way, it sets the tone, basing the interpretation upon given risk categories: so-called "natural" risks (for example, risks linked to earthquakes), technological risks (for example, risks linked to concentrations of hazardous materials), or sanitary risks (for example, risks linked to viruses).

There should therefore be a sequence upon which a hazard may be chronologically based. "The simplest" and "the easiest to understand" are, above all, the hazard, followed by the components of land occupancy or human settlement, including populations, exposed buildings, and even their unequal vulnerability, which are revealed by damage *a posteriori*. Indeed, damage necessarily comes after the hazard has occurred, according to "the individual nature of objects", if we resume Descartes's thinking.

Within the risk, which is the risk of damage, the hazard always seems to come first, and even gives the impression of being naturally foremost, especially within the category of "natural hazards". For example, it rains, part of the incident precipitation flows as overland flow, affecting the populations and buildings, and damage consequently occurs. Is not this development in and of itself "simple" or "easy to know" to move toward the issue of damage and finally risk, which means the potential future damage we should then better know and prevent? In addition, the sequence "assigns" an order, which "naturally" appears as essential, by seeming to instigate cause and effect relationships. However, it is primarily a question of an observation, which revolves around an evolution, which may then end up acknowledging the links between causes and effects. This is not necessarily the case. We observe here the essential nuance that Descartes introduces, by bringing attention to the difference between "assigning" and actually "demonstrating". The third principle is therefore much more nuanced than it appears to be upon initial examination.

The nuance is all the more essential if we acknowledge, continuing to quote Descartes, that good sense, which is to say, the "power of judging a right" or "reason", is "of all things among men, the most equally distributed". Indeed, but only upon initial examination, who would contest that damage may occur even without the rain which occurs "naturally" and, of course, first? This would equally follow for a riot, the H5N1 virus, a stock market crash or a revolution, which are all types of possible hazards. However, the Cartesian good sense, which aims at the most rationally coherent interpretation possible, cannot acknowledge that the damage may only be explained by the hazard, to use our example, only from the rain. This is proved by so many lessons learned, and also by so many previous studies. They show that damage is also – and even above all! – produced by virtue of the characteristics of human settlements even before the hazard is concrete. Take, for example, an earthquake of the same magnitude in different parts of the world. It has been a proven fact that the intensity of damage will differ with its occurrence, for example, at Port-au-Prince or Sendai. The hazard thus reveals the highly uneven respective predispositions of being damaged. It is not necessarily first, far from it. It therefore seems to us that what Descartes recommends in the principles relating to the Method, seeking both the simplest, easiest, and most coherent possible solutions in agreement with the principles of Cartesian "good sense", do not sit easily together. The simplest and easiest principle does not necessarily prove to be the most rational.

As a consequence, Descartes advocates here a methodological approach to "problems" which he claims is "the least limited" and, in particular, the "simplest" and "easiest to know". We are simply attracting attention to the fact that it alone may justify the thematic fragmentation seen across the different types of disaster risks. Based upon the fragmentation of definitions and problems both within simple sub-notions and categories, this systematic approach easily wins support. This is especially so as it introduces the fourth principle, which appears both more definitive and more useful. Descartes advocates the need to "...make enumerations so complete, and reviews so general, that [he] might be assured that nothing was omitted" [DES 37, p. 46].

The fourth method principle appears to prepare grounds not only for a systematic interpretation, but also for exhaustive purposes, useful for the problems which disaster prevention poses. Moreover, it is certainly this type of interpretation which continues to dominate numerous publications upon disaster risk. In attesting to this, among the thousands of articles and academic textbooks organized according this aspiration for systematic exhaustiveness, *The World Atlas of Natural Disasters Risks* [SHI 14] is one such book. It seeks to draw up a global inventory from the interpretation of the various subtypes of "natural hazards", its ninth part being devoted to a summary that reconstructs the concept in its entirety, that is "*understanding the spatial patterns of global natural disaster risk*".

As may be seen, the issue which these systematic interpretations pose is not only linked to the identification and the justification of "fragments" of difficulties or the so-called "simple" so as to be able to analyze a given "difficulty". Within the sphere of disaster risk prevention, the distinction between hazard, exposition and vulnerability is brought into general use, even if it is reductive, only leaving a meager part, which is very often implicit, to the policies themselves. The position granted to the scope or to the social status of these risks, which is differentiated according to the context, is also meager. There is also an issue posed by the relationships these "fragments" maintain, here between sub-notions associated with risk definitions. And worse still, all too often, the thinking surrounding risks is mainly reduced to considering such fragments separately: hazards, vulnerability, issues at stake. The Cartesian "good sense" should urge us far more to reconsider the dominant causal interpretation of both damage and risks from the hazard, which is highly reductive, and which continues to exist even though many deny it. The systematic and monocausal interpretation turns out to be insufficient. Not being content with it would assist us to better understand "the inescapable plurality of competing principles", the dynamics linked to disaster preparation throughout the duration, as well as the surprises which are associated with it. The nuances of Cartesian thought are evident: Descartes never wrote that man was the master and owner of nature, but that "knowing the force and actions of fire, water, air... allows us to become, as it were, masters and possessors of nature". The expression *as it were* stresses the limits. However, it is clear that the temptation for systematics, the fruit of the analytical fragmentation of issues, which Morin calls the "Major Disjunction", dominates [MOR 86, pp. 70–71]. It appears to us that we find these limits with the analytical fragmentation of the notion of risk, which is centered upon the sub-notions of hazard and exposition/vulnerability, and even more so when the interpretation is hazard-centered.

The systematic approach is attached to developing simple approaches, and facilitating understanding, but at the price of limitations. These same limitations allow us to better understand those which disaster prevention policies encounter. Responding to the challenges they give rise to implies the need both to pursue the systematic approach and to supplement it by using an alternative approach. This is what results from systemic analysis.

5.2. Reconstitution of the analytical foundation: Pascal and the systemic attempts

As such, Pascal cannot be described as a specialist in systematics, for fear of appearing anachronistic. A system explains the relationships, which are identified between several elements, which, at least in part, interact. The notion is considered to be a later one; it gradually appeared over the course of the 20th Century, courtesy of De Rosnay [DE 75, p. 91]. Significantly, De Rosnay also mentions that it is shared by several university disciplines. Moreover, he does not present it as something new. We might claim that Pascal was a precursor of this complementary way of understanding problems. Indeed, we may be struck by the similarities between the basic definitions of a system, that we may find, for example, in De Rosnay [DE 75], Walliser [WAL 77] or even basic definitions of complexity, such as that found in Morin [MOR 05] and some of Pascal's contemplations.

For De Rosnay [DE 75, p. 92]: "According to the most common definition 'a system is a set of interacting factors'".

Morin [MOR 05, p. 48] is very clear in saying the following:

> "What constitutes complexity? At first sight, it is a quantitative phenomenon, the extreme quantity of interactions and intrusions between a very large number of units...However the complexity does not only consist of quantities of units and interactions which defy our possibilities for calculation: complexity also consists of uncertainties, vagueness and random phenomena".

It seems possible to compare these definitions with this extract from Pascal, especially because Morin [MOR 11], in fact, invites it himself, by stressing the contribution of this 17th Century philosopher to the understanding of issues which affect us today. As Pascal [PAS 63, p. 527] said:

> "As all things are caused and causative, aided and aiding, mediate and immediate, and as they are mutually supportive through a natural and imperceptible tie that binds together the most distant and most varied things, I hold it impossible to know the parts if I do not know the whole, and impossible to know the whole if I do not know the particular parts" (*Pensées, dans Oeuvres Complètes*[2]).

The relevance of the systemic analysis for the set of themes around disaster risk prevention was acknowledged by large parts of the research undertaken, especially by those working upon socio-ecological systems [GUN 02]. For us, the similarities between these three fundamental definitions are very close despite the time intervals between their finding expression, thus indicating a very dated problem for humanity which consistently crops up. It revolves around knowing how to both think and act for the best, indeed acknowledging the need to do both but also their necessary limitations. Moreover, systematic analysis provides results, which, in isolation, are not the most satisfactory, if we acknowledge the significance of trying to find them, for example, by following Descartes. Furthermore, in this context, who could possibly question the need to prevent more disasters, especially in relation to ethics? This is particularly the case when hoping to reduce mortality rates or even economic losses.

---

2 This translates as "*Contemplations*, in Complete Works".

In this sense, the contributions of systemic analysis are multifaceted.

For example, the interactions between the elements of a system make it possible to better understand why the relationships between hazard and damage cannot be reduced to simple cause and effect relationships. This is particularly the case each time direct or indirect human abilities are actually influencing the hazards. The latter are not necessarily foremost within in-depth analysis, and they interact, at least in part, with other human settlement elements. To understand the dike risk and the policies which justify it, we can no longer consider the hazard as foremost, and we are obliged to acknowledge that the hazard interacts with both corrective and protective works, which are dikes, as a minimum. Indeed, the protective nature of corrective work should equally attract attention to the existence of the relationship with human settlement elements, which are supposed to be protected. We then return to the surprising discretion of risk management policies within definitions, as well as within conceptual models, which revolve around the prevention of disaster risks. However, there is no shortage of examples of corrective works, or techniques for ecological engineering, which all pursue the objective of modifying the hazard. How can we fully understand the mechanisms of a green roof, without taking account of, among other things, the will to struggle against this building's contribution to the potential formation of the urban heat island? Studies abound which tackle the urban heat island as an unintended hazard, which is partially advanced by urbanization and which may be rectified. Green roofs are a form of political initiative which intend to prevent the disasters associated with urban climate development. They can equally bring about interactions between urban areas (the exposition and vulnerability factor) and climate (the hazard factor), which call for numerous forms of political initiatives.

Furthermore, there are interactions between system elements, and even more interactions between several systems, which represent a form of complexity [DAU 13]. These also make it possible for us to better understand why it is impossible to find a simple and sufficiently interpretative statistical relationship, while being immediately useful for initiatives, between hazard and damage. Such a simple causal relationship may make it possible to predict any future damage. However, the research studies which have been conducted in association with insurance companies, for example, upon the relationships between the height or speed of water and the damage done, do not provide meaningful or usable results (see

[DOM 08, AND 13] and Figure 2.4). Again, this so-called operative dearth in respect of the causal link, despite the evidence of Cartesian simplicity, reveals many other hidden factors. There are various degrees of damage according to the building structure, building materials, building maintenance, current furniture and other factors. The shortcomings in information inferred from the single causal link between hazard and damage do not only result from the flaws in information collected by insurance companies. Once again, we can observe the impossibility of transferring an argument that is current in the sphere of physics toward that of disaster prevention. Arnaud says [ARN 98, p. 180] the following:

> "...gas state variables are not independent and, for a very long time (the 17th Century onwards) we have sought to make laws which translate their mutual dependency. We may break down the issue, taking these variables in pairs (the relationship between pressure and volume or between pressure and temperature and indeed other variables), with the two other variables remaining constant. This was the historical approach, which ended up with the formulation of a number of specific laws".

Within the disaster prevention sphere, despite recent attempts to return to this problem, concerning the quantitative relationship between specific hazard and damage [AND 13], there is no equivalent of the Boyle–Mariotte law, which quantifies the pressure–volume relationship.

Finally, systemic analysis makes it possible to bring to light the feedback loops. Such loops reconstruct simple causal links between the various elements, which form a system. By identifying a feedback loop, we also identify an interaction: that is the cause and effect, which seemed linked by a simple causal relationship, then interact, and become sub-elements of a system. To achieve this, it is necessary to identify that the effect returns to the cause and converts it. In this case, the initial presumed cause is, in return, converted by its effect.

Prevention, as such, assumes the possibility of a retroactive effect on a cause that is assumed to be the chief contributory factor explaining the damage previously experienced. This seems abstract, but we have numerous available examples of this type of process, set out simply as danger signs. This is shown in Figure 5.2.

Figure 5.2. *How a warning sign heralds systemic analysis – Groton, Connecticut, August 2009. Photo: P. Pigeon*

In Figure 5.2, potentially, the hazard is the assumed cause of social unrest. The hazard is the man who, in 2009, took the photograph in Groton, within a residential district, and who indeed is one of the authors of this book. In all of these cases, we are obliged to admit that without prior acknowledgment of the damage sustained, the sign, which brings about a political decision, may in fact not exist. However, the sign sends a very visible and explicit message which the eye illustrates. If you, the potential cause of injury, indeed become a real cause of injury, in short a hazard, do not forget the Groton community is watching you. As the potential hazard, you, who watches us, become vulnerable, given that we are also looking at you, and we also have the means to bring pressure on you should the need arise. This emphasizes that the initial assumed hazard, the cause, which makes others potentially vulnerable, then also itself becomes potentially vulnerable through the medium of potential victims. These victims themselves become potential hazards, and the sign is thus heralding this. The interaction calls into question the strict division between the hazard and the exposed element, which has varying degrees of vulnerability. We merely have three interacting elements, the basis of a system, if we take political decisions into account. The sign, which is dedicated to the prevention of a type of damage, in harmony with the decisions taken by a section of the Groton population, heralds the system. This policy assumes the existence of previous damage, and it seeks to anticipate this by using lessons learned, in reconsidering previous damage. The sign brings about at least one feedback loop. However, the representation remains fixed and static and, above all, implicit. As such, the sign is not a representation that is explicitly systemic of a political measure.

Out of necessity, systematic analysis is fixist and static, if not short-sighted and thus having the advantage of myopia, which makes it possible to achieve greater precision with reference to details, and also greater immediate effect. On the other hand, systemic analysis is more directed toward taking into account the length of given evolutions. It makes it possible to consider trends, ruptures and indeed, surprises. Who could not see how much use, and even how necessary systemic analysis proves to be, as soon as we claim to both understand and better manage the limits of disaster prevention policies? After all, such a challenge imposes the need to take the long-term perspective into account. It is exactly the same for those lessons learned. How can we envisage depicting a lesson learned by a simple and solely linear development? The logical limitations of this type of representation justify having recourse to systemic representations.

Systemic analysis may therefore provide greater logical coherence to both definitions and conceptual models of disaster risks. Thus, it also corresponds to Cartesian "good sense". However, for all that, it cannot claim to eliminate all limits, as much within the sphere of knowledge as management. Philosophers such as Bergson [BER 34a, BER 34b, BER 41], or more recently, Edgar Morin [MOR 82, MOR 05, MOR 11] or Dupuy [DUP 05], insist upon this point. Moreover, this is particularly the case when in-depth consideration revolves around prevention in the long term.

It seems to us that Bergson can assist us in understanding why all thinking which is directed toward dynamics and predictions is necessarily limited. The views of Bergson, but also those of Morin or of Dupuy, help us to identify and understand the impossibility of eliminating all damages, as well as all limitations to efforts of knowledge.

5.3. The limitations of analytical reconstitution: Bergson's intuition

Bergson insists a great deal within several of his works upon the limitations of understanding and upon its conditioning. Bergson says [BER 34a, p. 22], that actual understanding is "far from being a disinterested work".

Bergson believes, in particular, that limitations are the price to pay for being able to handle and understand situations. Indeed, he says, human intelligence is fundamentally directed toward the need to act, and as such sacrifices a "moving reality", which is somewhat unthinkable to given states, to "real or virtual stances", as so many approximations are just as necessary as imperfect for knowledge:

> "This substitution is necessary to common sense, to language, to practical life, and even [...] to positive science. Our intellect, when it follows its natural inclination, proceeds on the one hand by solid perceptions, and on the other, by stable conceptions" [BER 34a, p. 36].

Moreover, this inclination for human intelligence assumes to be practical selection and is orientated toward the acquisition of knowledge. Bergson says:

> "Long before philosophy and science existed, the role of intelligence was already to manufacture particular tools, and to guide our actions towards those around us. Science pushed the work of intelligence much further, but it did not change its direction. It aims, above all, to make us masters of the subject." [BER 34a, p. 76].

Bergson deduces that these two points (fragmentation of knowledge and pursuing mastery of the subject matter) have several limits, which support the previous views, such as those of Descartes and Pascal mentioned earlier. He stresses in particular "...the problems and contradictions which naturally result if intelligence is specifically intended for the study of a given part and if we nevertheless claim to use it, so as to gain knowledge of the whole" [BER 34a, p. 76]. We again encounter the division, which does not consist of opposition, between systematic analysis, focused upon the individual parts and systemic analysis, which is more directed toward "knowledge of the whole". Bergson stresses particularly well the non-exclusivity of these two approaches, which are divided up between the intuition of objects and of phenomena, which falls within duration, and an intelligence, which is displayed as a discontinuous flow, based on interruptions and states. This uneasiness nevertheless corresponds to significant difficulties for risk management, inasmuch as the sensitivity of understanding is disturbed by the duty to act. Besides, Bergson acknowledges that in practice "...our actions are only trained conveniently upon fixed points: it is therefore fixedness which our intelligence seeks" [BER 34a, p. 49]. Human intelligence captures imperfectly that which is both global and dynamic, while disaster prevention policies are, out of necessity, directed toward the obligation to both understand and handle as best as possible the global, which is necessarily multi-factorial, and takes account of dynamics.

For Bergson, it is thus a question of finding the means to fight against "natural" human intelligence trends to both separate and fix, so hopefully, to be able to both handle and understand. He states:

> "How should we eradicate such a profound inclination? How should we cause the human spirit to overturn the direction of its normal operation [...] to no longer only see, within decisions and states, snapshots taken across moving objects" [BER 34a, p. 112].

All of these different elements seem to us to shed light upon the extreme difficulties in finding conceptual models, having the greatest possible logical coherence, as we observed in Chapter 3. Moreover, this is the case despite the abundance of lessons learned which show, almost immediately, the limitations of existing models, on at least two grounds. Indeed, it is necessary to make a great abstract effort in the hope of systematically representing disaster preparation, and therefore also in the hope of contributing to their prevention. Still, we can only just understand a fraction of the systems that disaster preparation implies, and whatever the efforts toward their justification prove to be. Besides, the representation can only offer a possible and partial solution for an evolution which the analysis immobilizes. Definitively, Bergson himself acknowledges the current inclination of thought when he asserts that "...the philosopher will be simply lazily sitting back to deduce consequences according to the rules of a rectilinear logic" [BER 34a, p. 156]. However, this simplistic solution does not only apply for the philosopher – far from it!

We may deduce the significance of the following excerpt from Bergson's work regarding insurmountable contradictions with respect to knowledge development:

> "...within his eternally unquenched desire to espouse the object around which it is condemned to turn, the analysis endlessly multiplies given points of view to complete the representation which is still incomplete, which tirelessly varies symbols to perfect the translation which is still imperfect. It continues to infinity" [BER 34a, p. 57].

The analysis is therefore necessary, but irremediably limited, in particular when it is a question of capturing changes and evolutions. Moreover, recourse to systemic analysis no longer makes it possible to eliminate this fundamental problem, as "...analysis, by definition even, always leads to given states" [BER 34a, p. 81]. Now, "...a state, taken in and of itself, is a constantly changing entity" [BER 34a, p. 92]. Systemic analysis hopes to push back the limits of systematic analysis, but it cannot entirely eliminate them.

Bergson offers a solution which we will not tackle in this chapter, as here we are distinguishing the pathways shared between philosophy and efforts at understanding disaster knowledge and the disaster prevention policies that we are suggesting. Faced with the relative and analytical fixedness brought back to the measurement standard, comparison or analogy, being what we already know, Bergson contemplates the absolute, fluid movements and intuition. Indeed, Bergson says, "...the analysis takes effect upon the immobile, while intuition should be seen within the context of mobility, or that which boils down to the same thing, in terms of duration" [BER 34a, p. 93]. As intuition cannot be converted, translated, indeed represented as a symbol – at the risk of returning to within the field of analysis – we can understand why Bergson claims that it lies in the realms of metaphysics. Yet, Bergson says the latter "...isn't intended for any application..." [BER 34a, p. 114], and differs in any case from a work of knowledge which is far from being disinterested.

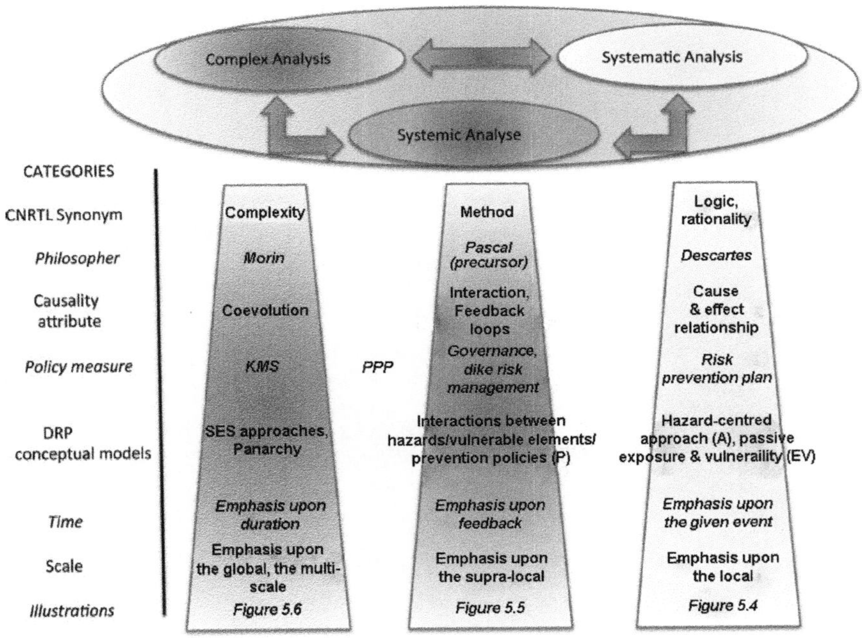

Figure 5.3. *Systematic, systemic, complexity: a diagram of analysis types. PPP = Public-Private Partnership, KMS = Knowledge Management System, SES = Socio-Ecological Systems, DRP = Disaster Risk Prevention*

We are thus condemned to think and act imperfectly, the additional element being that Bergson helps us to understand why; this results fundamentally, "naturally", from the characteristics of the human capacity to both think and act, which are, out of necessity, limited. Adjacent elements have been tackled by numerous philosophers, Morin among them [MOR 86], relating to the paradigm of complexity. He too denounces the mutilation of thought by analytical categories upon which systematic analysis is based. This mutilation may be alleviated through systemic analysis. However, we cannot entirely escape mutilation thinking, or knowledge conditioning, as soon as we intend for its immediate application. Disaster prevention policies, which are the issue here, do not belong to metaphysics as Bergson contemplated. The point remains to find a solution to the limits of the various forms of analysis, which are, in practice, often interplaying, whether the analysis is systematic, systemic or complex (Figures 5.3–5.6).

Figure 5.4. *Justification for systematic analysis: Le Tour, Chamonix (2002). Photo: P. Pigeon*

COMMENT.– *The hazard, in this case the torrential River Arve (A), threatens the exposed elements, which are vulnerable to unequal degrees (EV). Earth sciences study the hazard, human and social sciences study the degree of exposure and vulnerability. The analysis is systematic, the interpretation monocausal, with the hazard remaining of the foremost importance.*

Figure 5.5. *Justification for systemic analysis: Le Tour, Chamonix (2002). Photo: P. Pigeon*

COMMENT.– *The hazard, in this case the torrential River Arve (A), still threatens the exposed elements, which are vulnerable to unequal degrees (EV). This is why we also see on the same picture corrective/protective works (CP) : in this case, dikes. Dikes transform the hazard (river Arve) and are supposed to protect exposed elements. The corrective/ protective works contribute to justifying urbanization, see Figure 8.10 for chronological accuracy.*

However, the limits of the corrective/protective works are acknowledged by three generations of dikes : disaster prevention policies had to be reconsidered. In this case, it lead to have more dikes, but also more exposed elements, and to transform even more the river Arve itself. Hazard (A), exposed elements (EV) and corrective/protective works (CP) interact. The analysis is systemic, feedback loops appear and the hazard is no longer foremost. Please see Figures 8.9 and 8.10 for more details.

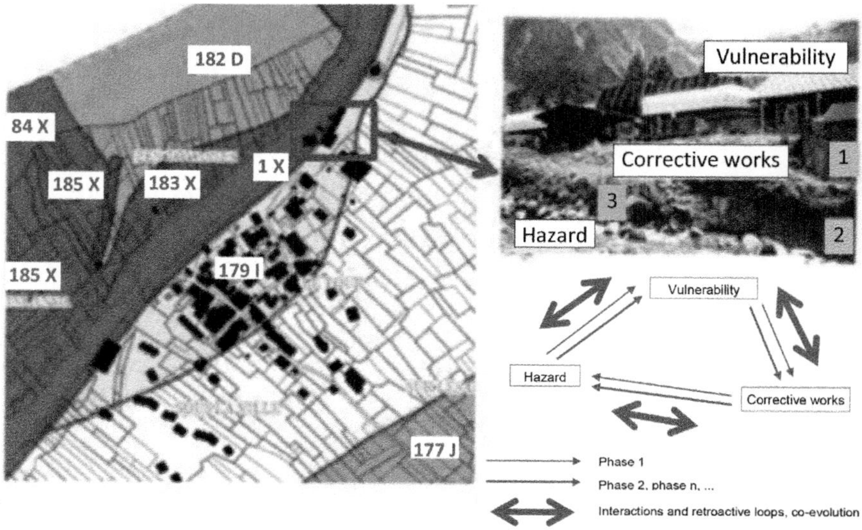

Figure 5.6. *Justification for complex analysis: Le Tour, Chamonix (2002). Photo: P. Pigeon.*

COMMENT.– *Despite three generations of the correction/protection process, there are zones which limit urbanization on the Chamonix Risk Prevention Plan (RPP).*

The latter, which originated at the state level, stressed that the issues linked to disaster prevention are not only local. The local land occupation system co-evolves with other land occupation systems on regional, national and international scales.

The Balme cable car, which provides access to one of the ski areas of the Chamonix ski station, highlights the international scale.

Disaster prevention policies should take into account the various issues at different scales, and associated local institutions and players.The analysis becomes complex, also necessitating partnerships between research and management; this justifies PPP and KMS/panarchy.

Following Morin's train of logical thought, the challenge presented may be reformulated, according to [MOR 05, p. 21], when he says:

> "Political strategy requires complex knowledge, as strategy is conducted by working both with and against the uncertain, chance, the multifaceted interplay of interactions and retroactive effects".

However, beyond these principles, how should we apply them in order to contribute to better preventing disasters while knowing that we will be, out of necessity, "naturally" limited in this undertaking? Furthermore, recall there are so many additional forms of conditionings besides those solely related to epistemological framings.

Epistemological limits to both absolute knowledge and certainties are both multifaceted and impassable. The fact that disaster prevention policies are confronted with numerous questions which do not have a satisfactory response (as explained in Chapter 1) is not thus in and of itself a problem. Always having increasingly precise knowledge is not the only condition required in order to act in a better way. It is not so much the lack of a perfect response, which impedes this. It is instead the difficulties in both understanding and acting in an imperfect and uncertain world, faced with a highly structured modernist and positive form of risk management thinking [VEY 05]. Rather than always trying to reduce the irreducible, and persisting in that within an impassable paradox, which intrinsically links both necessary and flawed prevention policies, it is about accepting the cognitive limits within development and implementation of disaster prevention. It is not a question of renouncing knowledge so as to act. However, it is also a question of recognizing that it is fruitless to claim the absolute or the ultimate explanation, where the limits of access to knowledge are impassable. It is, in addition, lessons learned which invariably show us the plurality of a moving reality, which defies analysis (a point which is particularly developed in Chapter 7).

Moreover, the cognitive limits do not only stem from epistemology. They have ramifications both for research and policies and always present policy implications. We should remember that Bergson stresses that intelligence and the understanding of the action to take also rely on both social and political contexts. It is impossible to separate them from such conditionings which are part of risk creation and indeed environmental issues across territories. Thus, disaster prevention confronted by its cognitive limits, and indeed the challenges which complexity poses, combines with other environmental issues and initiatives (such as sustainable development, climate change, adaptation strategies or the promotion of resilience), which are currently unavoidable in the field. This is developed in Chapter 6, which follows. Disaster prevention is certainly not an isolated concept across given territories. Besides, it closely interacts with a larger environmental issue as a broad societal matter, also wrestling with major uncertainties, both in terms of knowledge and action. This articulation adds both to difficulties, and also to opportunities to cast pertinent light upon disaster prevention, for which the concept of flaws does not necessarily prove to be a limitation.

6

Bridging the Gap between Disaster Prevention and Environmental Concerns

The paradox of disaster prevention policies results from numerous elements: limits linked to measurement, conceptual frameworks, analytical capacity to access knowledge, multifaceted conditioning of this knowledge and other similar factors. Furthermore, the intensification of risk concern and the proliferation of management initiatives are inseparable from an environmental issue which indeed proved to be a social issue in the second half of the 20th Century (see section 6.1). The context of such concerns makes the social and political conditioning easier to understand in various situations. Those conditionings affect the production of knowledge around risks and also the development of policies to prevent them.

Within this context, we may ask if taking account of multifaceted conditioning of knowledge around risks and disaster prevention will make it possible to extend the scope of questioning, and to exceed some of the cognitive limits, which are linked to both the specialization and analytical breakdown. Conversely, we may also wonder about the capacity of such a broadening of risk concerns and risk prevention matters toward environmental issues, to truly tackle the recurring obstacles to both thought and action (see section 6.2). Given the broadening towards environmental concerns, to what extent is it possible to bring to light aspects which have traditionally been neglected by analytical and segmented approaches? May we genuinely go beyond the recurring dissatisfaction that flawed actions necessarily inspire? May we move beyond readings that are reluctant to accept limits, which any management policy and disaster prevention understanding are bound to meet? When all is said and done, does embedding risk concerns and their prevention in the environmental issue constitute an advantage? Alternatively, is it perhaps a disadvantage? On the other hand, is the essential aspect of the challenge, which poses the need to act despite this shortcoming, in fact elsewhere?

6.1. The gradual contributions of environmental interpretations to disaster prevention

6.1.1. *Emergence and strengthening of environmental concerns*

The concern for disasters and their reduction is nowadays indissociable from broader environmental concerns. Upon reading the various academic works, the latter start to develop a structure as a social issue from the mid-twentieth Century. The increase in environmental awareness has contributed to this ever since the widely popularized message of scientists. This was the case with the work of Rachel Carson [CAR 62], or that leading to the Meadows report [MEA 72] through to the public opinions which are henceforth deeply concerned by the topic of climate change. The scientific literature also covers to a large extent environmental concerns, which are becoming both institutionalized, and international, and which progressively assume a global dimension, through, in particular, the topic of global warming.

International negotiation arenas are appearing under the aegis of the United Nations, such as the Conference on the Human Environment in Stockholm in 1972, or the 1992 Earth Summit in Rio. Although the Stockholm conference was the first to set out environmental issues in international terms and in conjunction with development, in 1992 the Earth Summit set up the notion of sustainable development and indeed created an action category relating thereto. As a consequence, the so-called Agenda 21 or sustainable development has become for France not just programmatic principles but integrated into local level initiatives. Such international networks are formed around sustainable development, as with the signatory cities and towns of the Aalborg Charter of European Cities and Towns toward Sustainable Development in 1994, which are committed to pursuing the principles of sustainability on their respective individual scales.

Within this international context, which has a major focus on environmental concerns, the United Nations inaugurated the decade for the reduction of natural hazards in 1990. In 1999, the initiative led to the implementation of an agency responsible for international strategy for the reduction of natural disasters (UNISDR –the United Nations Office for Disaster Risk Reduction). The agency worked on the adoption of the Hyogo Framework for Action covering the period 2005–2015, as well as the Sendai Framework for Disaster Risk Reduction for the current period (2015–2030). These documents bind signatory states to adopt disaster reduction measures in line with commitments given.

However, the issue of climate is, without doubt, that which occupies the majority of the global dimension of environmental issues. At the Earth Summit in 1992, the United Nations Framework Convention on Climate Change was adopted. It was within this framework that the Kyoto Protocol was established in 1997, which came into force in 2005. The latter binds signatory countries to both control and indeed reduce their greenhouse gas emissions. The conferences of signatory parties (or COP) constitute regular meetings of signatory countries to this framework convention, which makes it possible for countries to update their individual commitments. The climate issue justified the setting up of an Intergovernmental Panel on Climate Change (IPCC) in 1988, at the request of the United Nations Environment Program, and the World Meteorological Organization. Scientists have thus met to shed light on the political decisions around environmental issues which did not arise initially in terms of international relations. Dahan Dalmedico says that the issue of climate, which will soon become that of climate change then of global changes, powerfully demonstrates the extent to which the scientist and the politician, when considering environmental issues at the global scale in this case are difficult to separate [DAH 07].

6.1.2. *Environmental issues and society interaction*

The environmental issue may be posed in terms of a social issue based on the fact that it plays a significant role within opinion and public debate, that it is subject to an institutional context weighting and that it makes the positioning of collective initiatives inevitable, and also highlights the individual commitments regarding environmental issues. However, the environmental issue is also a social issue insofar as the given social context, in return, shapes part of the scope and characteristics of the environmental issue. The place, the time or the context, both social and political, in which the environmental issue is formulated, are not strangers to its main features.

Nevertheless, in this sense, as regards the social context, the environmental issue not only concerns physical processes, or hazards. It is also a product of historical facts and of society.

Global warming is a very good example of complex links, which inextricably link physical and societal phenomena, and which question, by their acknowledgment, the modern separation between nature and society, as Descola also states [DES 05]. David Demeritt [CAS 14] shows how global warming was scientifically constructed, in a selective (social) way at the end of the 1980s, as a global atmospheric physics issue. The solutions envisaged were conditioned by this initial format: the global scale and the physical origin of the failure (the issue of greenhouse gas emissions), placing emphasis on mitigation strategies. The development of scientific literature in the field encompasses how the climate issue has been transformed over time. It has gradually moved beyond the basis of global atmospheric assessments, on a scale which is closer to the given territory, in a more complex way that is more attentive to both the functioning of given societies and their relationship with their surroundings.

As a milestone of this evolution, the emergence of the notion of adaptation within the academic literature on climate (primarily as an analytical category, and then as a particular action category) comes to complete the notion of mitigation and targets local levels. Rebotier indeed states that adaptation places emphasis on the uneven reality of the phenomenon of climate change both across territories and among societies [REB 14].

Posing the environmental issue in terms of a social issue acknowledges that environmental issues are differentiated according to social sectors, players, territories or even scales implied. They manifest differently, but they are also composed differently, well beyond solely the physical processes (hazards) which define them. Thus, in a reciprocal way, the various social and historic conditionings contribute to defining the contours and the specificities of these selfsame environmental issues. The latter integrate both physical and social phenomena, as much within their realization as through their making. Beck states that acknowledging such complexity justifies calling into question modern divisions, and provides for developments across reconstitution of both modernity and hybridization [BEC 86]. In a constructivist perspective, we should indeed be attentive to each of these phenomena, so as not to naturalize an environmental issue among which drivers, as with consequences, always prove to be socially conditioned.

6.1.3. *The environment – the divide between science and politics*

Armed with this broader constructivist understanding, the works within social sciences underline some implications of the environmental issue, which are far removed from physical processes (hazards). However, those implications are highly significant, and even recurrent, among issues and obstacles encountered in the field. Thus, the attention directed to various types of social and political conditioning helps to contextualize some inherent limits and flaws of environmental policies. Indeed, the implementation of these policies presents a significant democratic challenge for public action, which the case study of the Grenelle Environment in France measures and the climate change adaptation strategies illustrates.

In 2009 and 2010, France adopted, in two phases, a legislative package known as the "Grenelle Environment Forum", which encouraged local authorities of more than 50,000 inhabitants to implement adaptation strategies in response to global warming. The second part of the Grenelle law in 2010 stipulated in particular that Climate-Energy Territorial Programs (CETP) should be adopted by 31 December 2012 at the latest. France's Environment Code was accordingly amended. It in particular states that the CETP are regulatory tools which are supposed to set "strategic and operational objectives for authorities so as to reduce global warming and adapt to it" (article L229-26 of the Environment Code). The emphasis is placed on local territories, bottom-up initiatives and the suitability of given policies, based around "living with" far rather than "fighting against". However, field studies have shown the numerous difficulties for local authorities to implement national adaptation initiatives, which are stimulated by the Grenelle Environment Forum, and sometimes also opportunities seized to develop territorial programs. These largely go further than the single issue of adapting to climate change [BER 13, SIM 15].

All too often, taking Aquitaine as an example, adaptation strategies developed by local authorities bear the hallmark of normative and burdensome regulatory frameworks. They only partially take into account related issues, which we may think at first glance are not related to adaptation or global warming, but which are, in fact, closely interconnected within a complex, territorial logic. Rebotier notes that adaptation strategies lastly demonstrate more traditional difficulties linked to top-down public action, where the state fixes the objective and remotely manages through Key Performance Indicators

[REB 14]. Adaptation policies have proven locally to have numerous limits (in the cases where they were effectively implemented).

For the smallest authorities, it is the lack of both human resources and budgetary means, which prevent the regulatory injunction taking on a dimension other than strictly administrative or technical (for example, regular formal workshops, basic organization of consultation mechanisms, development of good practice guides, diagnosis – with large consultancy fees – of greenhouse gas emissions). Simonet [SIM 14] says that there is little place for a transforming and introspective dimension for adaptation. Still, this context is more promising for significant changes, providing greater scope for the social dimension amongst the environmental issue, and it is in this way more prone to arouse support and appropriation.

A number of actors were questioned on the ground (politicians, local authority technicians, who are well prepared but swamped, and those working as cattle breeders and farmers in the Pyrenean mountain range). They assert that the implementation of adaptation strategies mainly happen without their input or knowledge, the decision-makers being ignorant of both their situation and their interests. This happens to the point that some are asking where indeed these initiatives may be developed first, as such initiatives are unfamiliar to them. There is a lack of understanding although the acknowledgment of the necessity to act is unanimous regarding global warming and the future disasters associated with it, as an abundant literature and a strong institutional backing clearly state. Some actors are indeed at a point where they intentionally do not support, even deliberately boycott, adaptation initiatives, which are perceived as too top-down and disconnected from the territory [REB 13a, REB 13b].

Moreover, the technical and regulatory composition of adaptation strategies most often falls within a systematic analytical approach to environmental issues, as was set out in Chapter 5, detached from the specificities and the complexity of the territories from which they arise. This approach, denounced as top-down or crushing by players on the ground, rests on an expert composition, upon the introduction of technical mechanisms, and upon the low visibility of other forms of conditioning [REG 15a, REG 15b]. This systematic, regulatory and top-down approach discredits other styles of discourse, other rationalities and ways of thinking, specifically because of their non-scientific nature. On the decision-making front, the concept is still very present, according to which better forecasts (of tomorrow's climate) help us to act better (politically today). This was the tenor of a speech given by a think tank member on climate change in Aquitaine[1], at the start of a public meeting held in 2013. It led to an inventory of climate knowledge within the region. The scientist concerned, a researcher in geology, said that scientific knowledge should serve as the basis for decision making. This is a prerequisite. It is necessary to technify the question so as not to make it appear partisan. The work of scientists then consists of simply saying, "Here is what may happen now, do as you wish".

During the implementation phase, this is a way of putting the burden of responsibility for the potential failure or the limits – which we know are inevitable! – on environmental policies. However, the experience on the ground would suggest that obstacles in support of environmental policies or the poor acceptance of flaws and the uncertainties of knowledge rest clearly on a modern interpretation, in line with Morin's "Major Disjunction" [MOR 86]. The dominant environmental approach rests on this interpretation, when seeking to base policy around science. In other words, the obstacles come less from the difficulty of drawing "good" lessons learned out of constantly improved knowledge (which we know are, out of necessity, limited), than from biased questioning which accounts for the plurality of conditioning of environmental issues, while there is a consensus upon the need to act. Despite this, we are left with the idea, taken from the world of Shadocks, that it is a matter of improving the inventory of knowledge to adopt better policies, while ensuring that we do not lead science astray by partisan logic.

The environmental issue is indeed a problem which the society poses directly to itself, as much as, within this dilemma, the prevention of disaster is fundamentally a political matter. A constructivist perspective of disasters, including their reduction, finds its place within these debates on the environmental issue. However, for all that, the possibility, through broadening the environmental issue, of advancing social and political conditioning of disasters and their prevention does not give more keys to respond to the paradox of the necessary politics, while at the same time proving to be flawed. Indeed, environmental policies seem to be dealing with questions in the manner of a technician, and are indeed very top-down,

---

1 The public meeting referred to took place on February 12, 2013 at Saint-André de Cubzac. The think tank's findings were published later [LE 13].

which brings them closer to disaster prevention policies carried out in the DRR field. This, necessarily attracts attention once more around key research questions, which have been tackled previously. However, this leads to exploring the advantages and disadvantages of the overall increase in environmental concerns in order to escape the recurring paradox around disaster prevention policies, which turn out to be as necessary as limited.

6.2. Advantages and disadvantages for the increase in complexity for disaster prevention

6.2.1. *Virtuous effects of risk inclusion within the environmental issue*

The broadening of the viewpoints of disasters allows us to depart from an interpretation focused on hazards, and indeed base disasters as social constructs, which also fall within social phenomena [GAR 05, RIB 10]. The constructivist interpretation of disasters is certainly dated [O'KE 76]. However, it nevertheless remains a current challenge to be understood. Indeed, if the Stockholm conference articulated environmental and development issues as far back as 1972, we are still faced with major difficulties when it is an issue today of articulating disaster prevention policies with multifaceted conditioning which, in addition to hazards, enters the realm of disaster creation. The diagnosis is set down: a better disaster reduction policy must be to stop talking in terms of disasters [LAV 13]. An improved policy both for housing or spatial planning consists of a policy of disaster reduction because it diminishes territorial vulnerability through selective attention being directed toward the so-called essential elements and issues at stake [D'ER 09].

The significance of the confusion of multifaceted drivers implied within disaster creation is nowadays largely acknowledged, so that the Millennium Development Goals, updated during the special summit on sustainable development in September 2015, were articulated by environmental concerns taking the form of Sustainable Development Goals.

The intricate connection between environmental issues and development is confirmed, on the one hand, by the links which exist between the promotion of sustainable development or responses to climate change, and, on the other hand, the reduction of disasters. Thus, resolution 45d of the

Sendai Framework for Disaster Risk Reduction (2015–2030) states that the implementation of the Framework objective undergoes integration "of disaster risk reduction measures into multilateral and bilateral development assistance programs, within and across all sectors, as appropriate, related to [...] adaptation to climate change [...]" (United Nations, 2015, pp. 25–26). The glossary of terms linked to disaster reduction that UNISDR promotes indicates that adaptation not only affects climatic factors, and that "many disaster risk reduction measures can directly contribute to better adaptation" [UNI 09, p. 4). Moreover, climate change is supposed to translate into an expected increase in disasters linked to extreme climatic events [IPC 12]. Henceforth, strategies for disaster reduction and those of adaptation to climate change are often linked within the framework of the promotion of resilience.

Morin claims that, to some extent, the notion of resilience may be linked to thinking around complexity, which makes it possible to consider the entanglement of multifaceted conditioning which, "woven together", has something of the nature of the creation, tensions and inconsistencies of the environmental issue [MOR 82]. The systemic analysis invites a break with modern understanding of environmental issues, which separates hazards and physical phenomena from societal phenomena, or which does not allow for the visibility of the plurality of interactions. This systemic analysis is therefore particularly adapted to the understanding and the formalization of both the hybrid and the future composite. De Rosnay says that it makes it particularly possible to leave the simple, linear and binary thinking, to take into account the pluri-factoriality, and the role of feedback [DE 75]. By stressing undesired effects in case flood defenses burst, public authorities acknowledge possible feedback of disaster prevention in France (and indeed, in this case, protection) through the dike risk (developed in section 4.2). The mobilization of systemic analysis to understand environmental issues and their complexity presents heuristic assets, which are potentially significant. From this point of view, the notion of panarchy indicates the spatial and temporal articulation of systems on different levels, and their various resiliences, which prove to be contradictory [GUN 02]. This mechanism, for example, makes possible the explanation of scalar contradictions for flood risk reduction issues, which preserve significant issues at stake of the regional urban structure in Lyon or Geneva. They maintain their resilience, to the detriment of other issues at stake which are considered to be of secondary importance being much more local, upstream. The resilience of municipal settlements is thus called into question, in particular by blocking urbanization programs on the major basins of the Rhône or the Arve [REB 13a, REB 13b].

Another advantage of the notion of resilience is that it attracts attention to the possibility of preventing disasters while being unable to eliminate all damage. Walker and Salt [WAL 06] give the definition: “the capacity of a system to maintain its essential structures in case of any disturbance”. The resilience does not thus designate the absence of damage or the possibility of eliminating all damage but the possibility of limiting the intensity of future damage “in case of any disturbance”. This is what makes it possible to explain that human settlements may be maintained in spite of the disruption. It is possible to attract attention to the coherence of this interpretation with disaster prevention. Indeed, the maintenance of essential settlement system structures makes it possible to limit the recourse to foreign aid in case of any damage. The converse is accepted as one of the four criteria making it possible to define a disaster, within the CRED (Centre for Research on the Epidemiology of Disasters) meaning, as we saw in Chapter 1. However, the researchers of the Resilience Alliance explicitly insisted on the fact that resilience is not a universal panacea, and that promoting it may contribute to preparing disasters as well [GUN 02].

Indeed, in parallel with its formal assets, the resilience world has recently become hegemonic within the production of contemporary academic works [NOB 14], regaining numerous gray areas with regard to links undertaken with adaptation or reducing disasters [GAI 10, KEL 15b, REG 15a].

Pigeon comments that, in this way, the recourse to systemic analysis also does not prove an end in itself, at risk of copying a functionalist viewpoint developed within ecology across social systems, as per the arguments of some supporters of socio-ecological systems analysis [PIG 16]. Nothing, from the viewpoint of system resilience, should allow for judging good or evil based on both the configurations and dynamics of social systems. In the eyes of the field’s leading authors, resilience is neither good or bad [GUN 02, p. 31]. The authors say that it may also translate in functional terms to: “…the ability of a given system to return to its original structure following a particular disruption”. It is possible to foretell the success of this notion within the sphere of disaster prevention, and also the danger of considering the return to nearly the same conditions that prepared a disaster [WIS 04]. Indeed, Pigeon observes by virtue of the so-called “rigidity trap”, that the example of La Faute-sur-Mer shows how the reduction in the frequency of low-intensity damage by diking flood-prone areas may contribute to maintaining the essential structure of a system. However, it can lead to population and human settlements dynamics behind the dikes eventually favoring the creation of disasters, which are both rare and of high damage intensity (2012). In this case, a functional interpretation of the resilience of the territorial system regarding disaster prevention proves limited, even simplistic, and calls for being surpassed within an operational perspective [WEI 14].

Systemic analysis makes it possible to identify multifaceted interactions (feedback) and types of conditioning, which play a part in the creation of the environmental issue, which is closely related to the risk construction perspective. However, complex thought, with the conceptual tools which it provides, potentially encourages us not to confine our analysis to an interpretation which is simply functionalist and decontextualized. The consideration of risks and their prevention thus benefits from cropping up more in terms of complex thought, which is less organized into a hierarchy and more relational than in terms relating to modern and linear thought, which both technifies and depoliticizes. However, employing resilience (in particular in [GUN 02], is an advantage in understanding but may simultaneously be a disadvantage with regard to action. There is thus constant tension between systemic conceptual frameworks giving sense to environmental issues, and more systematic frameworks for both functionalist and naturalized interpretations, which can have a tendency to be exploited.

6.2.2. *Confusion, diversion and exploitation: are these inevitable pitfalls of complexity?*

By advancing relational aspects or the multiplicity of factors, complex thinking is distinguished from modern and systematic thinking which separates, fragments, specializes and interprets within a more linear causal relationship (see Figure 5.2). However, this propensity toward the whole and to articulation may also produce a genuine black box of the confusion (including from a scalar perspective) of diverse conditionings, which poses more problems than it offers in terms of solutions. The view, which is widely spread, of articulation between adaptation to climate change and disaster reduction finds numerous objections within the academic debate. For some, the integration of climate change themes, sustainable development and disaster prevention is evidence of efficiency [SUM 10] while for others, both the understanding and the surpassing of particular conflicts and contradictions of these various environmental agendas across territories would first make it possible to make progress [SOR 11]. Worse, the advancement of the climate change lexicon would be a decline in understanding of (and in initiatives upon) social and political conditioning of such disasters [QUE 13]. Finally, some, for example, Mercer, question the innovative scope for articulating these notions [MER 10], while, on the ground, the compatibility of the operational agenda of each of these does not happen without posing problems [RIV 14].

Moreover, the obscurity of themes has become extremely technical, having something of the nature of their confiscation through ranges of expert discourse and, as a consequence, their simultaneous depolitization. The complexity, which is valued from a heuristic viewpoint, gives way to complication thus muddling analytical work. The still larger technical nature of environmental issues would suggest that it suffices to resume physical sciences or earth sciences to both reduce uncertainty and increase decision-making efficiency [HUL 09, FOR 03]. Within this inclination for the scientific method, the reduction in flaws of environmental policies involves sciences and technology, as the only way to grasp formulation of issues posed. The expectations are excessive not only toward modern science or expertise in earth sciences, but also toward engineering, which dominates technology, or the industrial sector which currently prioritizes technological development. They leave out of account political and social dimensions, which however condition the characteristics of these phenomena.

This dual black box impact of complexity (confusion and complication) may hinder more than serve as a means for understanding the creation of environmental problems. For some, it is even the means of exploitation. The modern fusion of scientism and capitalism [GID 98, LAT 94] is in this sense increased. The integration of environmental issues into social, economic or political dimensions (although formulated on an international scale as long ago as the 1972 Stockholm conference) is reduced to more compartmentalized and systematic arguments of both science and modern technology. We seek less to act flawlessly by placing environmental policies, both socially and politically, than to deepen knowledge of hazards, on which our action should be based. Thus, the social scope for sustainable development is progressively neutralized by a scientistic, technical and regulatory interpretation [DAH 13], close to the observation which may be drawn up of the most current recent usage of the notion of adaptation to climate change within the relevant specialized scientific literature [BAS 13].

Within the record of this confiscation of consideration through technology or the scientistic logic, the functionalist and normative use of resilience may give additional leverage to modern, linear and binary thought around the environment [FEL 14]. Even more so, Walker and Cooper [WAL 11] see within this use of resilience an ideologically oriented form of exploitation. The authors locate the proximity between, on the one hand, the complexity in ecology, worked upon by Holling [HOL 73], and, on the other hand, Van Hayek's neo-liberal economic philosophy. Both viewpoints, in particular, reject the idea of the stable equilibrium of the complex (and social) system. Besides, we find here the interpretation of the constant dynamics of "creative destruction", which has already been indicated, which hints at forms of capitalism [SCH 42]. For Walker and Cooper [WAL 11], the application of resilience to social systems in this sense serves a political agenda. The interpretation of the growing success of the greenhouse gas market, among which the technical nature both defuses and then, *a priori*, neutralizes all forms of politization [FEL 15], or even the observation of strong market development linked to technical responses to environmental challenges [JAN 14] has nothing further to add to the matter.

Possible drifts of conceptualization of the environmental question may contribute to the confusion, the complication and then the confiscation, and even toward a form of instrumental diversion to ideological and economic ends. It thus becomes difficult to even acknowledge virtues (even heuristic ones) of complex thinking around the environment. However, it is the classic dichotomy of modern thought, which is more linear and binary, and more systematic, which both desocializes and fragments knowledge around risks (we come full circle to Morin's "Major Disjunction" [MOR 86]). In cutting all ties with social contingencies, the positive epistemology (being modern, more linear and binary) only explains narrow phenomena; these are "states" within a moving reality. It is in breaking with this bias and fragmentation, based on limited and disciplinary certainties, that complex thought may possibly contribute to thinking differently about the limits and irreducible flaws of environmental policies.

Thus, in favor of relational thought, which makes way for the entanglement of multifaceted conditioning, for flaws and contradictions of situations studied [CHA 14], our approach is less one of reducing uncertainty or seeking perfection than acknowledging the plurality of rationalities and multifaceted interactions, which are often contradictory, that characterize the disaster creation. The epistemological constriction of an approach which is particularly linear and causal more than it tends to be relational and complex, which is always exposed to possible diversions, is certainly not the least of the difficulties to overcome.

In this sense, the reflexiveness seems to constitute a beneficial course of action, both for realizing the complexity of situations studied, the elements which are "interwoven with each other" and their multifaceted feedback; both returning to methods of knowledge production, and the endless questioning of the line of thinking around risks and the environment, which is all too often taken for granted.

Disaster prevention, not only its limits and contradictions, but also successes and social issues that it regains, takes on another dimension within a larger environmental issue. The articulation with the invasive issue of climate change certainly makes possible the interlinking with multifaceted factors which intervene within the risk creation, and which count within the development and implementation of disaster prevention policies. However, although disaster prevention thus falls more easily within an entanglement of factors, its study and the understanding of the paradox which cause tension between policies, which are both necessary and necessarily limited, do not, as an automatic consequence, progress. Indeed, the challenges which are posed are similar. This, nowadays, gives a sense of the fact of linking research on climate change to those of disaster prevention. This evolution is strongly supported by international institutions. The Antalya forum mentioned above, and held in February 2015, thus gathered together both researchers and managers who specialized in the field of climate change adaptation and others working in disaster risk reduction. It was sponsored by the Global Meteorological Organization, and also by the International Red Cross and the World Bank.

The increase in complexity, which accompanies the environmental issue, equally might put forward certain processes, as much as it might dissimulate them. Complex thinking does not necessarily assume its own conditioning. The recourse to reflexiveness may be envisaged as people distancing themselves and as a reaction to a modern and linear perspective, separating elements of complexity, to adopt a more relational interpretation to both the development of disasters and their prevention.

The articulation of disaster prevention in the context of larger environmental issues, and, in particular, the thinking around complexity marked by reflexiveness, offers valuable stimuli to produce original knowledge. Despite this, we cannot necessarily succeed in breaking with the ways of thinking, powerful narratives, or dominant paradigms that condition thinking, which emulate from the so-called "Environmental Social Sciences" movement [CAS 14]. Faced with the paradox that postulates the need for disaster prevention policies, which are both necessary and always flawed, social sciences would be able to elaborate a specific viewpoint, on the basis of a peculiar epistemological world, and the purpose of which is the social sphere. Producing such knowledge allows us not only to examine social aspects of given issues (with the sub-issue of how these are defined) which the environment places on societies, but also to study the given environmental challenges for particular societies [LÖV 15]. The general objective is indeed that of completing different forms of risk analysis and prevention thereof, in which fragmentation is a recurrent obstacle. Furthermore, the issue of compartmentalization of knowledge types also arises for geographical knowledge, whether "physical" or "human", which are produced around given risks. The complementarity of forms of analysis and knowledge produced is the subject of the following chapter.

7

Reflections upon the Contribution of Social Geography to Disaster Prevention

"As a result of the diverse representations, although material reality is one, its knowledge is multifaceted and nothing *a priori* guarantees cross-cohesion…Even if each researcher may defend his preferred model's given rationale, we should not forget the diversity of world views, and thus its potential models…and reduce geography to a dominant disciplinary ideology" [BAI 97, p. 21].

In its ambition to understand "the diversity of geographical systems" which are located upon "the Earth's surface", geography struggles to find its place within the modern classification between earth sciences and social sciences. It resorts both to given skills and to argumentative motives within these two spheres to span the multiplicity of the "processes which explain this differentiation"[1] (upon which we cannot take an exhaustive approach, as Chapter 5 reminds us). Human geography, for example, aims to explain global differentiation through social relationships and relationships which societies maintain with their given environments as Bailly and Béguin state [BAI 82]. Among such correlations and relationships, social geography, as a branch of human geography, places the emphasis on "aspects of societal life" which contribute, in various forms, to both "global differentiation and spatial organization" [BAI 82, p. 101].

Regarding environmental problems, and particularly disaster prevention, technical and hazard-centered approaches continue to dominate the

1 http://www.hypergeo.eu/spip.php?article386, consulted on 2 December 2015.

epistemological world in terms of knowledge development [PRO 15a, PRO 15b, HUL 09]. This is true despite an acknowledgment of the essential openness to other differentiation drivers and of the significance of enabling dialog around various interpretations of environmental issues [JAN 06, CAS 14]. Epistemological avenues are suggested so as to break with the model of modern and specialized science, and move toward that of more hybrid forms [BEC 86, LAT 94] or indeed to cultivate a more systemic relational analysis.

In this chapter, we propose to explore both the opportunities and the pitfalls that resorting to social geography within a constructivist epistemology offer us. Faced with it, other epistemologies and lines of argumentation are simultaneously at work to both understand and manage disaster risk. In particular, we think about experimental epistemology, which exposes the analysis to the potential for refutation, through experience or the argumentative motives of earth sciences. In the latter, the restrictive definition of given research subjects does not allow for the consideration of social rationales. Our work will initially revolve around putting forward the two major approaches, which are traditionally distinguished between both physical logic and, of course, human and social rationales (and of which the dividing line is nowadays widely called into question in critiques upon modernity – see section 7.1). Second, we will see how it is essential to put into context the range of constructivist approaches in social sciences around both the environment and risks, in particular regarding their "social utility", which is often set down as a pragmatic form of injunction (section 7.2). Finally, this chapter will discuss specific contributions of the social sciences, and in particular social geography, to the environmental issue and to disaster prevention (see section 7.3).

7.1. What epistemologies are necessary to develop environmental knowledge?

The concept of epistemology relates to a system of reference for "rational procedures by which a given theory is organized". Thus, this makes it possible to indicate whether argumentation generated within this system of reference "responds effectively to the models of any specific recognized scientific legitimacy" [BAI 82, p. 32]. Although the "scientific styles" differ among the various social sciences, we believe that it is guaranteed that they will always devote themselves to produce given understandings of the world, which turn out to be shared, inasmuch as they fall within a rational argumentative reasoning [PAS 96]. Given reasoning and methods may vary within the practices and traditions of the diverse social sciences. We do not find any less a form of unity in this, at least for some of these practices and traditions, having the ambition to develop a contextualized understanding of the world.

In line with the limits raised by Bergson (in Chapter 5), it is not a question of revealing "the" absolute sense of things, but of recognizing the possibility of knowledge on many different levels, which is simultaneously valid. Equally, we seek to agree a social reality intelligence, which is set in motion by both approximations and images, multiplied and incessantly subject to field tests. Envisaging this multiplicity in the form of a "shifting reality" within the complex fabric of risks and their "appropriate" management, it involves revealing the extent of the challenges posed for both knowledge and action. These philosophies are difficult to lay down, either as part of functionalist or technical approaches or because the risk subject (and its management) is framed by the given disaster or hazard. However, unlike philosophy, constructivist research in social sciences is confronted with both requirements (and limits) for knowledge as much as the necessity to act.

7.1.1. *Various ways of framing and various subjects*

The epistemological world of sciences is on a plural footing. Several systems of reference coexist and make scientific discourse concerning the environment possible, which do not fall within the same system as regards production of evidence upon such matters. Within this plural world, the conceptual framework for earth sciences turns out to be more unitary, marked by the definite and logical characteristics, arising from the positivity which has constituted them. On the other hand, the conceptual framework which constructivist social sciences belong to, corresponds to a field of coherence that is far more diversified than the rational field of the so-called exact sciences. Passeron says that fields of coherence are consistent with the "finite world of definitions and operations defining the syntactic and semantic constraints of any language which is capable of either never saying more or something other than it asserts on the world" [PAS 95]. Thus, the dominant language of the coherence field in an epistemological world open to constructivist approaches and sensitive to pluri-factoriality is far less marked by logical forms (for example, mathematical constructs) than the language of the coherence field of a more positive and realist epistemological world. Such a world should be based on the refutation raised by an experimental reasoning, and on seeking causal relationships.

In this sense, social sciences which do not claim a form of positivity do not produce a discourse upon things *per se* but upon things just as they belong to a context, over the course of history. Their description of the world is never only a description, assumed to be neutral and objective, which is, besides, independent of "the living mobility of things" [BER 34a, p. 42]. It is also interpretation, being the product of a view which is in tune with the world. In this sense, we acknowledge the weight of "situated", contextualized knowledges (according to Haraway [HAR 88]), on the one hand, and we move away from the universal and totalizing perspective in a scientific view of the world, on the other hand, which aims to reveal laws governing its functioning (only to draw out causal relationships). The environment is concerned with constructivist social sciences, not as part of their physical mechanisms or their respective types of impact. This is not their purpose. It is concerned about them as part of what constitutes the environment among societies the organization and functioning of societies, which is therein implicit, and the way in which societies both consider and interact with it. The topics of questioning are different.

Thus, work in social sciences does not start with data handling produced as part of other epistemologies, or according to other topics of questioning. It starts with the design of a data collection system and the definition of an autonomous area of questioning. The construction of research subjects falls within the scientific work, with earth sciences, just as with human and social sciences. Such an emergence of a subject, within an autonomous, coherent and consistent field of social sciences, simply indicates that several interpretations of the same issue may coexist (subject to, for their scientific quality, both the consistency and the capacity to understand within a given coherence field). The subject of flooding may be constructed as a question of hydrological assessments, or as the result of land occupation within a particular area. Different "flooding" issues, with a scientific basis, do not give rise to the same methods, or to the same information-gathering processes, and neither to the same perspective for responding to the problems which flooding poses.

In this sense, Berdoulay claims that the same word may correspond to different research objectives, which refers to epistemological frameworks, interpretation protocols and a theoretical language, which may indeed be very different [BER 88]. The weight of evidence of disasters within the study of risks has, for a long time, been a major point in favor of better definite understanding of the hazard which seems to be of prime importance, and as inflicting damage upon societies. There is nothing more equivocal than thrusting forward the argument that "the facts speak for themselves" whereas the categorization of data, their selective collection, or even the question previously asked, frame, the given perspective, and thereby produce in part the very same facts. "The power of common sense, which only retains the assumed objective meaning (with notions such as a risk, an event or a given landscape), by denying its eminently subjective dimension, saturated with historicity, singularly complicates the construction of the given subject – at least, if this epistemological obstacle is not identified" [COA 06, p. 24]. An initial issue resides within the autonomy of the structure of research subjects to the extent that this structure implies either considering or obliterating various conditionings relating to these subjects.

7.1.2. *Different systems for collecting evidence*

For social geography, scientific controversy is not of a cognitive and positivist nature, but of an axiological and comprehensive one [ALD 06, RUM 10]. It is not a matter of understanding the world of things in a more precise way (for example, by refining climate change prediction models to know future average temperatures with greater certainty). On the other hand, it is an issue of interpreting and reaching shared understandings on the environmental concern, nowadays acknowledged as a societal issue. It is indeed an issue, for example, of exploring what uncertain average temperatures mean for society, the investment to which they are subject, the reasons why they are central – or indeed not – within the climate change issue. Even the social responses that uncertainties about temperatures imply may come into play. The factual measurement of temperatures and their evolution is not pointless. However, it is equally important to consider what makes this data collection and related measurements significant. In this way, we reveal a system of values, choices and arbitration, which contribute to differentiating the world socially (well beyond factual temperature measurements). The purpose is different. In this sense, it is the axiological and comprehensive world in which the world of things takes shape and direction, which is under scrutiny.

In an epistemological world of constructivist social sciences, forms of evidence-gathering do not, in fact, follow an accurately formalized argumentative process in which all stages succeed one another in a linear way. It is instead a matter of a complex approach, which is reflexive at each stage of knowledge construction, which consists of operating a back-and-forth motion between enunciation – regarding the general scope – and observation – which proves to be empirical [PAS 95]. In this, we discover the proliferation of images suggested by Bergson so as to better approach the issue from the perspective of a moving reality [BER 34a, p. 9], which is difficult to capture by systematic analyses, for which formalism often obliges restrictive division of the area under scrutiny. The reflexive process is no stranger to systematic approaches when, for example, earth sciences test their results again and again through experimentation (as the arrows between the various forms of analysis show, in Figure 5.3, which are illustrated in Figures 5.4–5.6). However, there is always a certain tension between detailed but incomplete knowledge of phenomena, and their comprehension which is more complete but superficially done. Sometimes, such findings are too indecisive to determine the one-off action, which is often sought, with respect to these same phenomena. For social sciences, for which the epistemological world is less marked by mathematical formalism and more open to common language, the constant trial and error research process is crucial. It entails possible actions including rectification, explanation and reframing. Moreover, it is through the dialog with practices and fieldwork analysis of other researchers that a field of coherence and assertion is created. This is a necessary step for the scientific validity of knowledge and views produced. Thus, social sciences do not lay themselves open in the way of earth sciences to more formal refuting of theories [POP 73]. They are no less legitimate to produce shared assertions across the social world, while they fall within a method of different knowledge production ([ALD 06], citing Habermas).

Thus, with the proof process, there is a complete change in vocabulary. Social sciences investigate more than they experiment; they interpret meanings more than revealing laws. They multiply counter-examples to orient interpretations more than they seek to refute a theory by experience, in order to formalize an updated result [PAS 95]. Without the premise of universality (at least nowadays [BEC 94]), which mixes the positivity of the sciences, social sciences as bearers of complex, reflexive and contextualized knowledge are not organized in general in the same way as earth sciences, which rely on more definite and positivist approaches. Although they are engaged in dialogs in their practice, it remains possible to distinguish these two major approaches, which justify different methods of knowledge production that prove to be complementary (Table 7.1).

	Complex, introspective and situated approach	↔	**Modern, positive and linear approach**
Prevailing argumentative format	Interpretation and analogy	↔	Necessity and proof
Favored methodology	Iterative process (between the given case and a more general scope	↔	Linear process (inductive or counter-factual)
Means of introducing written evidence	Increase in empirical observations compared with a theoretical discourse	↔	Refutability test for a given case from the viewpoint of a theory
Main epistemological system of reference	Plural mental space (emphasis on the continuous, less segmented aspects) ⇒ Recourse to multiple feedbacks, possible contradictions	↔	Formal discreet space (emphasis on the discontinuous, more segmented) ⇒ Isolating causality relationships)
Prevailing language	Everyday language Search for meaning Clarified polysemy	↔	Formal language (mathematical codification) Search for "truth"
Prevailing discourse	Plural intelligibility Producing meaning	↔	Tends towards the "truth"
Status of the debate	Axiological values and choices	↔	Metrological measurements and devices

Table 7.1. *Summary of the main epistemological, methodological or semantical elements peculiar to two major epistemological (and penetrable) worlds*

In this sense, the production of knowledge by social sciences is therefore less the result of a linear process than through constantly updated analogies. Experimental and positive epistemologies which support the search for results are different from epistemologies based on an iterative heuristic mechanism, which supports the contextualized production of intelligibility. The apparent rigidness of the initial reasoning contrasts markedly with the flexibility of the second reasoning. We stress, once again, that we may find rather "rigid" approaches and an ambition for the scientific character owing to the potential for refutation also among social sciences, which seeks to neutralize, at the most, the importance of values. Equally, we may find more "flexible" forms of reasoning, which are open to diversity, and the coexistence of alternative explanations among the earth sciences (or engineering sciences, for which uncertainty occupies, for example, an always more significant position).

Consequently, before it constitutes a disciplinary-based problem, the issue of interdisciplinarity raises the need for a common epistemological world, as a prior condition in order that the contributions of various disciplines may engage in a dialog. At the crossroads of these epistemological frameworks, there is a form of competition around the legitimate explanation between the "natural order and social order" of events [BAI 82, p. 30]. This issue, which crosses the epistemology of geography and occupies the more contemporary thought around reflexive modernity [BEC 86, GID 98], nowadays poses numerous difficulties, as much within research as within prevention policies. The conditioning relating to this issue (developed in the following section) contributes to explaining why research and policies revolving around disaster prevention continue to also be tightly fragmented and specialized. This also explains why approaches which are inspiring more systematic analyses continue to dominate the field.

A material reality, among those that concern us as part of environmental issues, is never just simply a material reality. It is also a historic reality. Being unaware that knowledge of material realities is linked to knowledge of historic realities is indeed a problem. It is not simply a question of acknowledging the area of coherence of social sciences alongside earth sciences. It is rather a matter of combining the skills to examine jointly, by co-constructing a shared area of coherence, the links that combine material realities and historic realities, and which indeed govern the characteristics of disasters, policies developed for them, and more largely the environmental question.

Part of the research activity tries to promote this shared area of coherence, for example, as part of the thematic networks of CNRS, as that around risk. One of the authors of this book ran this particular research project between 2005 and 2009. The willingness for pluri-disciplinarity within research does actually exist, and a proportion of researchers vote for it. However, difficulties start with the need to communicate using both notions and methods, which are able to be shared, for example, between geophysicists, quantitative economists and sociologists. Indeed, the pluri-disciplinary nature poses issues which predict those which are far more sensitive and delicate, within the realm of transdisciplinarity. It is possible to go from a systematic analysis alone to a systemic analysis as part of a given discipline. On the other hand, both the shared and potentially pluri-disciplinary acknowledgment of complexity make it highly tempting to go from one discipline to another. But such a tempting possibility is not without posing major epistemological problems (of which we have just highlighted certain aspects within this section). One of the major difficulties comes from the fact that it is simply a matter of challenging, given disciplinary specifics, and questioning epistemological frameworks, which support given approaches. Despite Morin mentioning that disciplinary division amounts to the death of thought, disciplinary thinking still exists today in 2016. It is, however, accepted that such thinking explains in large part the limits that research and policies devoted to the prevention of disasters encounter.

Yet, far more often, within the continuity of modern approaches which cultivate the division of knowledge, we demand that social sciences respond to the limits which earth sciences identify. These are also, for many, the limits of their own epistemology. Within these conditions, it is impossible for these disciplines to account for the historical and social dimension of the realities described. As part of interdisciplinarity, social sciences have come out of their comfort zone, called to meet at a point outside of their epistemological framework. Nevertheless, it is at this point that we often expect the most from them.

7.2. Under what conditions do social sciences contribute to disaster prevention?

The calls for both transversality or interdisciplinarity increase in response to urgent needs for efficiency when faced with the challenges that the environment poses to societies. The convergence of dispersed research efforts nevertheless poses a number of questions and cannot be thought of outside of the significant background in which the injunction that research upon the environment should be socially useful operates. This social demand is deployed within a broader societal context and a political economy. These in return precisely condition the forms of convergence and the articulation of different scientific approaches around the environment. This section proposes to deconstruct the loop of conditioning, which controls, through the demand for social utility, scientific contributions around the environmental issue.

7.2.1. *Interdisciplinarity? Complex epistemological combinations*

Documents from international organizations attest that, as much around the study of climate change as around risks, the approaches which were initially focused around physical aspects always seemed, more recently, to be more open to social aspects. The overall scientific approach is even attentive to the integration of the contributions to earth sciences, or indeed life sciences, with those of social and human sciences [JAN 06, ADG 09]. Having said that (and given the developments in section 7.1), it can seem difficult to articulate approaches and even scientific cultures. This is because the epistemological frameworks, the interpretation protocols or even the argumentative reasoning may be very different. Nevertheless, unquestionably, forms of comparisons operate.

They operate, for example, around keywords (such as adaptation, resilience, vulnerability, factors and strategies, to give examples of some buzzwords). They may even run through the sharing of grand explanatory narratives of the world, as with through the functionalist discourse on complex systems. The most fruitful collaborations between earth sciences and social sciences concern the following fields: economics, psychology or even political sciences [CAS 14]. When studied in detail, these comparisons turn out to be selective, and conditioned to respect basic principles, which characterize both the practice of earth sciences and a given scientific paradigm. They are especially marked by the utilitarian willingness to prevent disasters more. They have the necessary action in a given situation as their only outlook.

The study of the use of the notion of adaptation within the scientific literature, through the examination of four specialist reviews on climate change between 1996 and 2011, advances the weight of an instrumental and functionalist meaning of the notion. In 70% of articles concerned, adaptation is used as a simple form of adjustment of systems to climate changes [BAS 13]. Moreover, a study of public policies and mechanisms for participation linked to the climate issue in the United Kingdom between 2000 and 2010 shows how the epistemological, functionalist and utilitarian system of reference frames all of these initiatives [PAL 13]. Ways of thinking are dominated by the positivist paradigm, by the measurement tool and the strength of the methodology to guarantee conditions for refuting theories through new experiments, as evidence of the scientific quality. For David Demeritt, this prior framing operates within geography, when it is a question of researching the environment: "typically, the 'upstream' physical geography sets the factual parameters for some sort of 'what if' scenario, whose impact can be assessed by human geographers" [DEM 09, p. 128].

Eventually, this selective comparison between earth science and social and human sciences becomes exclusive of the specific aspects of which social and human sciences are the bearers. It sacrifices the deepening of plural understanding of environmental issues for a superficial arrangement, often with an operation use, but with limited scope for the production of knowledge [HUL 09]. Holling says that this tropism is accelerated by capturing the scientific debate on the climate issue, which conveys the acceptance of the notions of adaptation or vulnerability that have more recently been drawn from ecological functionalism [HOL 73]. However, within the history of the discipline, geography has pioneered the disciplinary relevance of the terms "influence", "adjustment" and then "adaptation" to understand the relationships between "man and the environment", a relationship studied since human ecology began in the 1920s [ALE 13, REG 15b]. The integration of scientific approaches configured by a functionalist and systemic ecological paradigm prevails in the context of the pressing climatic issue. It marginalizes the specific lines of questioning around social and human dimensions of environmental issues. Even more, the lack of critical reflection around this process of reductive integration contributes to the performative utterances of the environmental analysis, which in turn affects both social order and human settlement structures. This failure in terms of political analysis is pointed out, for example, in most of the critical literature on the notion of resilience and its use within environmental sciences [NOB 14], indeed precisely within the community for disaster reduction [MER 10, KEL 08, KEL 15].

The use of the same words, the same tools or methods and the same approaches for "further mutual understanding" is in no way trivial. The terms carry an epistemological framework from which valid assertions are expressed (while others, it seems, are consequently invalidated). However, concurrently, this framework couches the terms used for performative concepts, which enable the existence of a given reality, that make particular assertions, or even some questions, possible to the detriment of others. It is as much likewise for resilience and adaptation as for complex systems or stakeholders. In other words, the language of some epistemological areas colonizes other frameworks, by imposing expectations which turn out to be utterly contingent, under the guise of better communication and strength of the positive scientific position [PAS 96].

It is particularly this that Hakim Bourfouka and Nicolas Krautberger [BOU 12] show in their interpretation of an environmental history work which was written by Diana Davis appeared in the United States in 2007. The two authors discuss the status of history, its practice and its epistemological characteristics both through their experience of working in the field of environmental history and in Davis's work. In particular, they stress that the assertions which make it possible to describe, past accounts of theories on decline around the environment as instrumental rest on a positivist, instrumental and technical framework. The central argument of the 2007 book largely resorted to paleo-environmental sources to "re-establish" the previous dominant understanding of the issue. Owing to physical phenomena, which the technology allows us to understand, historical facts are reconsidered. Yet, it is indeed the province of history to restore – in a different way – what makes sense, within a given context, to understand the sequences and the succession of social realities that were correctly judged as "valid" at the time. The disqualification of historical sources around the environment, under the pretext of their biased character, by resorting to sources positively drawn from the physical environment changes the subject of the research. It definitely amounts to an "epistemological retrograde step" for history as indeed for social science [BOU 12]. Such a practice of environmental history exposes the discipline to the danger of a systematic registration within a positivist and experimental epistemology.

The difficulties of interdisciplinarity are not strangers to the dilemma, which arises from the idea that to be convincing and even "useful" to cope with the social challenges which the environment poses, social sciences should be "scientific". Thus, we understand that it is necessary to follow the canons of a positive system of reference, as a guarantee of objectivity and of authoritative knowledge to make the given decision. The "so-called scientific nature always consists in merging the rational area of the formal sciences, mathematics or indeed mathematical logic and, more frequently, that of entirely experimental sciences or the purely statistical reasoning, with the argumentative area in which the historical sciences actually reason" [PAS 95]. Within these conditions, it is difficult not to sacrifice the specifics of human and social sciences within the exercise of interdisciplinarity. This is all the more so since epistemological isolation, which is so prevalent within academic practice, is also conditioned by the pressing demand of "social utility".

7.2.2. *Social utility – the trap of legitimate framing*

When it is an issue of disaster-based research, it is difficult to think of another objective than reducing the number, intensity and impacts of such disasters. In this sense, the apparent paradox between the increasing number of researches and the assessments which would not better disaster situations to a significant extent, and even might worsen them, becomes of concern. The term "social utility" is then very swiftly convened, all the more so when it is a matter of public research, to assess the significance of particular results, and the advantage that can be taken of such an investment. It is perhaps not unanimous to think of "social utility" in these terms, but the social demand which emerges, and which reflects specific contexts, mirroring opinions [BLO 98], inevitably frames research, at least in part. Thus, in the United States, risk analyses are closer to a legitimate model for cost–benefit evaluations [GIL 13]. Other examples come from Latin America, and, in particular, from Central America, at the turn of the 1990s, where the social construction of risks was put forward within a context which was more sensitive to inequalities and development issues [MAS 93]. Scientific approaches and social expectations do certainly combine. Within society, sciences are thus confronted with demand for social utility. On this basis, they turn out to be eminently conditioned. More than complying with the injunction for utility, social science work also consists of deconstructing what this utility implies, what values and expectations it resorts to, or even what authoritarian narrative of the relationship between man and his environment underlies this.

The modern and technical approach, which is more linear and binary, which prevails within both the research on the environment and environmental policies, isolates the utility of both knowledge and action within a simple alternative [WHI 01]. On the one hand, existing knowledge would not be sufficient. Then, improving and completing research is useful to reduce uncertainties and produce a more appropriate knowledge, which is even more accurate or more relevant. More research is needed with a view to better action. On the other hand, existing knowledge would be sufficient, but might turn out to be either poorly understood or poorly applied. It is incumbent, in this case, upon social sciences to ensure the articulation between the production of technical and scientific know-how, and the implementation of appropriate policies. Shove says that it is the superficial ABC (*Attitude, Behavior, Choice*) logic in which environmental social sciences are often isolated, and against which it is so difficult to grapple [SHO 10]. In this regard, the utility of social sciences consists of securing the acceptance of scientific, technical and expert knowledge by actors on the ground. Those actors might be too far removed from such knowledge or may be unable (or even unwilling) to take it into account (supposedly whether because of ignorance, lack of competence, or selfishness, or other reasons [AIT 10]).

Social sciences are often confined within an epistemological framework, which fits other sciences or technical approaches, though not necessarily social sciences. Such epistemological bias in producing analysis around the environmental issue, and in our case, around disaster prevention, had already been pointed out in works within the field [LÖV 15, CAS 14]. The utility of social sciences is not then so much in their capacity to remove social objections, which would hinder the application of knowledge which becomes ever more specialized. It rather lies within their capacity not to keep (vainly) addressing the insurmountable paradox, which puts together both necessary and imperfect disaster prevention policies and research, but to think away from it. Nevertheless, its recurrent calling into question continuously runs into the same obstacles and limits, whatever their forms or terminology may be. We then tackle the crux of this work: the social injunction (indeed whose injunction?) being itself ethically and economically legitimate to improve prevention policies urges a functionalist interpretation and is assumed to be integrated. Paradoxically, this reproduces conditions that favor future disasters. Social sciences may be useful by ceasing to target policy imperfection as an issue to be eliminated. Imperfection then takes the status of an inherent quality of both the environment and risk management. This imperfection does not prevent the particular efficiency of this management, if it is acknowledged, discussed and shared. Certainly, the imperfection of prevention policies may always be reduced. It is also mostly important to understand that imperfection is contextualized and differentiated, anchored as it is within a society at a given moment in its history. This bears the hallmark of a given moment, of a place and of social relationships, which are necessary to account for, as there are so many possible ways to improve the effectiveness of prevention policies.

Weichselgartner and Pigeon [WEI 15] question the status of knowledge both in research and in action around the issue of disasters and their prevention. The authors point out that the status of knowledge is only rarely dealt with in the literature and poorly addressed in terms of action (where it appears reasonably well naturalized, as spontaneous and taken for granted). They go on to ask how we should think of disasters in a world of fragmented and plural knowledge, which does not necessarily turn out to be complementary, indeed which is sometimes in competition. Dealing with knowledge not as the sum of fragmented knowledge, but viewing it as a co-construction between various actors, and as a contested theme, is to make knowledge an eminently political and socially sensitive issue according to a given context (social, political, institutional or indeed based on other factors).

However, faced with this calling into question of a modern and linear epistemology, directed toward technology and practice, there are major challenges at the time of envisaging action, threatening significant expectations of social utility. Should we necessarily sacrifice contemplation and understanding, in favor of practice and action? On the one hand, are there not alternatives to tension caused that are often posed as impassable between simple and simplistic thinking of disaster prevention, but which makes action possible, and on the other hand the existence of more complex and qualified thinking, which accommodates inherent uncertainties and contradictions, though at the expense of the operational implementation of this knowledge, thus greatly complicated?

Social utility is not confined to utilitarianism with a consequentialist system of reference. Besides "doing", we might also "ask better questions" and be socially useful. In this sense, social sciences are useful, with a view to thinking differently about prevention policies, which are built around the illusion of getting rid of their limits. However, to be able to think in terms of this type of contribution, it is necessary to fall within an epistemological system of reference which acknowledges "the utility" (both for thinking and acting) not to find a solution to an impassable paradox, but to express differently both the status of imperfection and its relationships with science and politics. Yet scientific approaches argue, Shove says, about the monopoly of the legitimacy through their capacity to respond to social demand, not only by observing it rigorously, but also by shaping it and reinforcing it in turn [SHO 10]. There is competition for the monopoly of socially coherent explanations, which opens up the acknowledgment of markets, prestige, power and indeed other aspects. Within this competition between different research projects around the environment, the positive paradigm of simple and causal thinking prevails [MET 15]. It is thus difficult to defend any given "utility" outside of this given area.

The various types of conditioning through social demand are clearly related to the effects of institutional framing and political economics, which influence the development of knowledge around environment and risks. Epistemological and political drivers that take on a form of subordination of social science approaches turn out to be directly linked to these types of institutional and economic conditioning.

7.2.3. Conditioning through institutional organization and political economics

"The decisions to listen, read, argue with others, collaborate or recycle (or not do so) which are peculiar to professional researchers are most often those of so-called 'staunch nationalists' of a particular discipline. Indeed the chauvinism of this discipline is only understood by the demand for a monopoly within the legitimate research exercise which seems only to be able to be safely defined within a territorially constructed jurisdiction" [PAS 96]. The dialog between the various points of view never takes place outside of disciplinary power relations, indeed even beyond cliques of scientific researchers. These power relations between different scientific points of view are deployed as part of a political economy, which corresponds to the way in which a society is organized, regarding resources, allocation of values and profit redistribution.

As part of the main lines of contemporary research, particularly around the environment, certain approaches, questions and methods are in a dominant position. This dominance occurs through attracting grant finance, securing positions, legitimacy and the social prestige which is granted to them. David Demeritt [DAV 00] describes the evolution of the conditions of knowledge development, and the consequences around knowledge itself. He does so in the context of liberal reforms implemented in the United States and the United Kingdom when the "major political and ideological turnaround" in the final decades of the 20th Century took place [SUP 10]. He, in particular, describes the change in objectives that research is pursuing, and the societal frameworks within which such research is taking place. It is also a question of practices which are always more accepted through the canons of management, through the most immediate utilitarianism, and through cost rationalization, standing as so many cardinal objectives.

The development conditions for scientific knowledge upon climate fit competitive demands (regarding their results and social expectations). Such conditions may also respond to partnerships (with the entrepreneurial sector, as is the case for the financing of European Union research). In both France and in Europe, "we now only speak of innovation or of research driven by downstream industries"[2]. "Downstream" may correspond to the results "social utility" calls for. These are highly directed toward operational functions and the given moment; or even economic opportunities the research may lead to, as it is explicitly the case for numerous institutions and also research-funding agencies [DEM 00]. Faced with this strong polarity which is exerted upon research, some insist that "maintaining a research network, which we are in the process of losing, is fundamental. In France, there is increasingly less money within the labs and it is very difficult for young researchers to set in motion a research program – nowadays they spend their time looking for research financing options. At a European level, the ERC (European Research Council) program for financial assistance is too elitist".

The research conditioning is partly epistemological (as explored in the previous sections of this chapter). This conditioning does, however, combine with the constraints of a political economy concerning the status of researchers, research finance, the systematic resort to call for research proposals, as well as research evaluation and other factors. We are now more broadly in the presence of a major free-market narrative, which moves some values up to the detriment of others. The globalization that we are presently seeing attaches too much importance to quantification, measurement, assessment and technical know-how [YOU 06]. Moreover, the functional and instrumental dimension which distinguishes the use of the number of buzzwords around the environment is highly denounced in critical works in the field [EVA 14, GRO 14]. Furthermore, mercantile drift in response to the environmental challenges (which was touched upon in section 2.4) is a recurrent point within critical works in the field around climate [JAN 14].

2 This quotation and the following are taken from an interview with Jean Jouzel, published in the magazine CAES (the Action Committee for Social Mutual Aid), part of the CNRS (the National Centre for Scientific Research) – August 2015.

The resonance effect between science and politics is certainly nothing new, from the constitution of national science academies (representing the science fields) through to the free-market reforms which David Demeritt mentions. He describes the successive transition from science driven first by the search for the "truth", then by that of results (channeled by given states), then subsequently by that of profit (channeled by partnerships and the private sector [DEM 00]). There is therefore a power relationship within the clash between several axiological systems of reference around given viewpoints and the principles of hegemonic research around the environmental issue. The international institutions where numerous research agendas, deadlines and priority themes are settled constitute ideal platforms to account for the tensions to which the research around the environment is subject [REV 11, BOU 15]. During major international conferences on disaster prevention, and notably the Sendai Framework in 2015, some authors have stressed the pre-eminence of a technological paradigm and the significant position left open for both naturalist and technician approaches. This happens without thinking further of the diversity of perspectives around the fabric and prevention of disasters. Furthermore, in particular, no additional thought is given to the supplementary enrichment of various scientific practices, which produce different knowledge around disasters. The trend emerging from the Sendai Framework, produced from continuous negotiation, is marked for some by a decline in solidarity and social approaches, in favor of technology and expertise [QUE 15, REG 15b].

Thus, the obstacles to interdisciplinary work are numerous, and pressure from social utility is not necessarily the right compass to use to depart from it. It is not made clear whether or not an interdisciplinary frame allows achieving more accurate knowledge than the various disciplines do. However, different knowledge is thus developed. It does not suffice to share a coherent epistemological space, nor to establish a methodology which makes it possible to articulate various skills [REB 12]. It is also necessary to set and share the significance of an interdisciplinary task according to principles and ethical choices [COU 13], and thus break with uncritical alignment across dominant disciplinary line of questioning.

However, this positioning task, the increase in ethics of scientific practice around an environmental issue seized by the climate issue, is trapped in a vicious circle. Vaguely responding to the invitation to be "socially useful", as part of the technical and operational expectations, amounts to confine concerns within the trap of a legitimate but conditioned injunction. This injunction is blind to the vocation of part of the research (at least, the public research); that is to say, producing knowledge, criticizing world views and transmitting, if not disseminating, these contents and cognitive tools. We find here the three points developed in the following section.

"Climate is everyone's business!" but we are, moreover, led to ask whether it should be that of all scientific disciplines. Certainly, we could say it is, but not in an indiscriminate fashion. There is a tacit order among the sciences, which serves as a vehicle for this appeal which is set out as unanimous. The same provision applies to geography, a science of crossroads, which may be best situated to study the environment. Demeritt [DEM 09] shows us how to progress in the direction of this both urgent (and flattering!) invitation. He stresses the necessity, for the discipline, to take a critical distance from the viewpoint of underlying expectations, which both frame its task and its specific potential contributions.

7.3. What contributions do social sciences make to both disaster prevention and acting in an imperfect world?

Bringing together knowledge and action is not the main objective of social sciences (this is rather that of engineering, or industry, through technology), even if geography may contribute to this, particularly through the "development" pathways within its university curricula. In view of the recurrent paradox which we are concerned with, it seems instead important to know the effect of the environment on societies, what is said of it, the way in which it is commented on, or the place it occupies within social differentiation and human settlements structures [LÖV 15]. Besides the more technical contributions of geography as a discipline which aims at preventing disasters, social sciences may also contribute to charting the social fabric of disaster, not taking at face value the ruling expression of issues at stake. Social sciences may also raise the axiological challenge which consists of always bearing in mind the effect of risks and disaster prevention upon the social world.

7.3.1. *Charting the social fabric of disasters*

Research around risks, regarding the one stemming from human ecology, so far, has carried out studies on the relationships between societies and environments. A more radical school of thought is applied to put forward "root causes", which explain in part the differentiation of risk conditions among societies [OKE 76, BLA 87]. This work around "vulnerability" has contributed to denaturalizing risks and places emphasis around their social "fabric", and on place-based approaches. Yet within the more contemporary production of knowledge around risks and climate change, principles of the social construction of risks are marginalized as a result of the numerous types of conditioning evoked earlier, and which contribute to put forward responses based on technical, operational and functional frameworks.

Within climate research which started from the 1980s, the weight attached to physical mechanisms turned out to be a determining factor [ADG 06]. The mobilization of systemic approaches and complex thinking, attuned around the works in functional ecology from the turn of the 1970s [HOL 73], accompanied the circulation of common notions for both analysis and action. Within this epistemological filiation appear notions such as "thresholds", "factors", "feedbacks", "adaptation" or even "resilience" [CAS 14, p. 764]. These specifically lack the work undertaken around social contexts and how they articulate with risk creation or climate issues. The process of climate research is thereby different from that of risk research, especially when considered from the viewpoint of human ecology, which precisely addresses the relationship between society and the environment [BAR 23].

Social sciences should resist epistemological domination and hegemonic conditioning, to which the production of knowledge around the environment is subject. They should assert a conceptual autonomy, and defend their given objectives and the relevance of their line of questioning, without seeking protection from a refutable theory or a positive methodology. "Indeed, more than natural sciences and life sciences where technical application makes most of the differences in influence, prestige and profitability (whether social, political or economic) between disciplines, within social sciences the regime is different. The application of results is uncertain and, above all, it is hard to be evaluated by a state which brings funds. In that case, what is used as the best weapon of social sciences is the technical know-how and the esoteric nature of their formulations" [PAS 96]. By blindly forging ahead, we do a disservice to the specific nature of social sciences. There is no point for social sciences to occupy the scientific ground of the controversy around the specific subject of climate, or indeed, the measurement of climate changes. Social sciences involve formulating specific lines of questioning the social and territorial dimensions of the issues posed by climate to society, allowing displaying social conditionings. It is necessary to restore the plurality of the social world which shapes the fabric of situations involving not only the risk, the management solutions developed, but also the paradox of action and imperfection which arises as an urgent limit.

7.3.2. *A critical reflection around the expression of an issue*

The contribution of social sciences is not simply an issue of charting the social fabric of risks, but also relies on their ability to take some distance from the main lines of questioning both usually (and uncritically) acknowledged, from the most pressing points to other more neglected ones. In other words, the contribution of social sciences also consists of addressing historical conditions of the fabric of risks or climate change as a societal issue, which are necessarily differentiated within time and space, and among given societies.

Indeed, we know how conceptual framing matters and shapes interpretations, in particular when notions with a fluctuating or disputed definition are deployed, as in respect of vulnerability [OBR 07]. It is easy to determine the area (both material and conceptual) for a line of questioning around risks within a naturalistic perspective, and then to establish a posteriori a social sciences approach. This leads us to look into the social and human aspects, without thinking that strictly geometrical at-risk areas are not necessarily, in themselves, the same as broader "risk areas" [REG 15a, REG 15b, REG 15c]. Indeed, when the so-called "risk areas" are involved, the risks imply both area and scales, which greatly exceed the given at-risk area directly affected by a particular hazard. Not responding to a question posed by others (despite the pressure of social utility, or the temptation to take advantage of the legitimacy such activity comes with) is one thing. It is indeed another to question drivers and the scope for success of particular lines of questioning rather than others, or to wonder what institutions or what actors are promoting these, and with what effects on the social world.

It is impossible to be blind to the clandestine economic or political passengers conveyed by the use of one particular notion rather than another. Even more so, the use of the same notion, for example resilience, is sometimes the mark of a reproduction, and even a deepening of social relationships which are peculiar to a (neo)liberal political economy [WAL 11]. It is sometimes the opportunity for a reflexive and complex rupture with simple and linear thinking which invites consideration of both ethical questions and accountability within the sphere of governance [CHA 14].

It is by contextualizing, socially and politically, the frameworks for specific lines of questioning, but also the competition which drives them and their involvement in the social world, that social sciences help us to differently read and understand what the environmental issue and its management for society mean. That is what, for example, works in the field around resilience are seeking to do [SUD 14, REB 13a, REB 13b, ALE 13, WEI 14]. However, the uses and meanings that dominate both action and research appear to be highly different [BAS 13, PAL 13].

In this perspective, there are obstacles to the implementation of more efficient disaster prevention policies and to get people to buy into it. Not only are those obstacles linked to the content of policies or the conditions for their application (which reduce the problem to the famous ABC mnemonic which Shove devised [SHO 10], but it also becomes possible to contemplate "obstacles" significantly upstream of the complex process which leads to prevention policies. We are then able to address both its limits and potentialities on the basis of what makes a given policy appropriate or unacceptable (and, of course, for whom and on what scale does the policy apply?). The values, which serve to assess it or even the circumstances for creating it at a particular moment, are also of concern. These elements also make it possible to shift the emphasis of the questioning spotlight around the imperfect application of policies, or around their limited scope, to clarify their social and political articulation (the bias or the selectiveness of their scope, the differentiations which they implement within the given area or among the agendas of diverse players).

Social sciences are thus devoted to advancing social conditioning, which relates to the environmental issue. Their utilitarian scope is in this sense indeed limited, as it is not a matter of responding to the problems posed in a practical and direct way, but quite often to ask different questions (or at the very least other questions than the usual and dominant ones). The latter contribute to unlocking unknown aspects and underlying obstacles, making them both visible and explicit. More than within the uniqueness of a universal epistemological system of reference, it is within the plurality of interpretations that social sciences produce knowledge around the social world. Such knowledge is renewed and refined each time there is a new observation, or a case that questions the more general scope of an idea. This scientific approach seems to mirror more accurately the complexity and the diversity of situations observed, as well as the terms of the paradox which runs throughout this work. Indeed, for its part, a positive thought drawing on a rigorous and formalized logic can only partially capture it.

7.3.3. *Raising the axiological challenge*

Forsyth [FOR 12] states that among social science contributions to environmental challenges, the growing importance of political stakes, the acknowledgment of the plurality of situations and conditioning may be found. These are highly present within the works in this field. The framings and methods used, questions posed and the choices operated involve values, interests and choices. Putting forward axiological aspects in disaster prevention policies calls into question the naturalization of some interpretations, or terms which seem to be self-evident. The so-called "politization" of disaster prevention policies consists of articulating them to social drivers which are both numerous and often contradictory. As they are part of the making of disaster prevention policies, those social drivers also contribute to clarify the sense of disaster policies, their scope, implications or even limits.

Faced with this challenge which consists of scientific questioning falling into the axiological controversy [RUM 10], we sometimes find the prerequisite of axiological neutrality necessary for the scientific nature of any given assertion. However, "axiological neutrality" [WEB 63] does not mean the rejection of ideology, or the negation of sought-after values [KAL 05]. This principle does not assume stark objectivity, without compromising one's conscience, but acknowledging a given inevitable axiological arbitrary nature, and the relevance of its neutralization, if not its identification, within the argumentative mechanism. We have to be aware of the axiological world in which scientific assertions are expressed (as much around what is described as around the conditions of particular terms). Thus, it is not a matter of rejecting all values as inappropriate in view of the canons of science, but acknowledging their weight in the complex fabric of the social world and its interpretation (including by the sciences!).

Beside this political and ethical dimension, which concerns collective norms, their creation and what they imply for individuals features significantly a "missing link" (which is easily acknowledged!) in the epistemological references for resilience stemming from ecological works [GUN 02, p. 31]. The use of the notion of resilience within a functional perspective of complex systems assumes nothing as respects the conditions for choice and values, between actual scales and players. Nevertheless, this point is essential in explaining the conflicts which some prevention tools give rise to. If we succeed in understanding "how resilient systems operate", we will certainly not respond, through this epistemological system of reference, to questions such as "what" to do, and above all "for what purpose" and for "whom"?

Some within critical works in the field advance the performative dimension of scientific terms or public policies which indeed contribute to create a world of social connections or human settlements structures which convey values that have never been discussed [FEL 15, EVA 14, WAL 11]. Others elaborate from some values the typical ideal horizons from which given observed situations may be interpreted (see Fainstein [FAI 15] who formulates justice as a basic concept for assessing resilience). The definition of this system of reference undergoes explicit axiological choices. These choices may even turn out to be components of scientific practice, for those who feel drawn toward political ecology in particular [ROB 04], or for those drawn toward a critical social geography where "choices" then constitute research "principles" [ALD 06].

Refusing the researcher the possibility to place both his interpretations and his epistemological system of reference within an axiological world conveys the illusion that a scientific exchange of views may be neutral and objective, for which the single positive scientific nature indicates superiority. It also ignores that scientific discourse (even that of earth sciences) is never made "purely" available, freed from any contextual conditioning. Consequently, it is as if the scientist were refused "access to the political agora", even in the practice of his research [COR 11]. Forsyth states that as with all social activity, science is also political [FOR 12] and contemplates the differentiation of the world in which it is produced and indeed produces.

It is necessary, for all that, to beware of acknowledging the axiological dimension when contemplating disaster prevention. Taking account of this dimension assumes values brought into play, inasmuch as it is not necessarily a positive choice (creating some values rather than others, finding the "right", or the "true" ones), but rather a methodological component. Thus, the shifting of expectations or obstacles to prevention policies toward moral philosophy (favoring the "right" behavior), the collective slippage toward individuals or placing the emphasis around the problem on more local scales may shift the burden of responsibility (even guilt) closer to the territory concerned and indeed to the people at risk [RUF 11, QUE 13, REB 15]. This type of approach differs from global and collective perspectives. It thus contributes to exacerbating opposition to disaster prevention policies.

The mantra "the better informed we are, the wiser our decisions" does certainly apply, but only to a certain extent. One of the purposes of social sciences is to place this saying in social, political and axiological terms. Restricting the improvement of prevention policies to reducing uncertainties (having a better understanding) and their conditions for application (to better act) without questioning the nature, scope or social sense of them brings nothing new to the paradox of public policies, which are both necessary and imperfect at one and the same time.

The legitimacy of a dominant positive scientific system of reference, social demand or even political economics in the development of knowledge (set out in section 7.2) defines the significant pressure for the specific contributions to social sciences. Despite that, the contribution of social sciences concerns as much the social fabric of disasters, as the conditions for the development of knowledge, or as exposing axiological and political dimensions, which are always held in low esteem when contemplating the practice of disaster prevention.

The main purpose of social sciences is to reveal the multifaceted conditionings, which contribute to explaining the limits that disaster prevention policies encounter. They are therefore well placed to reconsider the paradox which is currently acknowledged, according to which the more we know, the more we are inclined to manage and, yet, the more we lose. Nevertheless, this purpose is only very partially compatible with those which earth sciences or engineering sciences pursue, for epistemological reasons which we have mentioned here, and which make it so difficult to envisage the complementary nature of knowledge generated. This makes it possible to understand why the injunctions for pluri-disciplinarity, and, all the more so, for interdisciplinarity, which nevertheless seem legitimate when faced with complexity, encounter as many problems in the academic world as within management practice. Within interdisciplinary programs around disaster prevention, social sciences often lose the aspects, which create, above all, their specific nature, being both the critical perspective and the revelation of conditioning. Yet, these concealed conditionings contribute to produce resistance and hostility, which disaster prevention policies both encounter and which also restrict them.

Social sciences do not, nevertheless, only exist to reveal the multifaceted conditionings which bear down upon disaster prevention policies (and which contribute to explaining their limits). The last chapter illustrates how these contributions by social sciences come into operation, thus contributing to both thinking and acting in an imperfect world. Despite the numerous limits indicated, social sciences are not compelled to resort to Bergson's intuition nor to its metaphysics only, with which they are often mixed up.

8

How Can We Best Manage Disasters?

Extrapolating this question further, we may ask whether it is possible to make a practical contribution to the prevention of disasters all the while taking into account the multiple and often structural conditionings that weigh on policies to do so. It has already been asked by several researchers working in conjunction with managers and policy makers. That is why we begin this chapter by addressing the practical component of the answer, which tends to have its roots in knowledge management systems (section 8.1). These systems (knowledge management systems, or KMS) are support tools for the assessment of disaster prevention policies, as well as decision-making support tools. However, they remain highly contested (section 8.2). Despite the continued progress in this field, they are only able to partially integrate knowledge from local players, such as citizens or households, who are likely to often challenge disaster prevention policies, as they are perceived as coming from above ("top down"). As such, local players in fact denounce part of the conditionings that frame those policies.

Figure 8.1. *An example of a mismatch between prevention policies and implementation of policies. Magland, Haute-Savoie, in 2000. Both houses are located in the floodplain of the Arve, and also in the blue area of the Risk Prevention Plan of the town of Magland (Haute-Savoie).The owner of the building in the foreground took into account the architectural requirements in line with the RPP and raised the living space. The owner of the building in the background has not taken the same requirements into account*

Figure 8.2. *Official SLLRDC mapping of flooded areas in 1992. SLLRDC Archives not classified (in Pigeon [PIG 08, p.223]). The rectangle locates Figure 8.3*

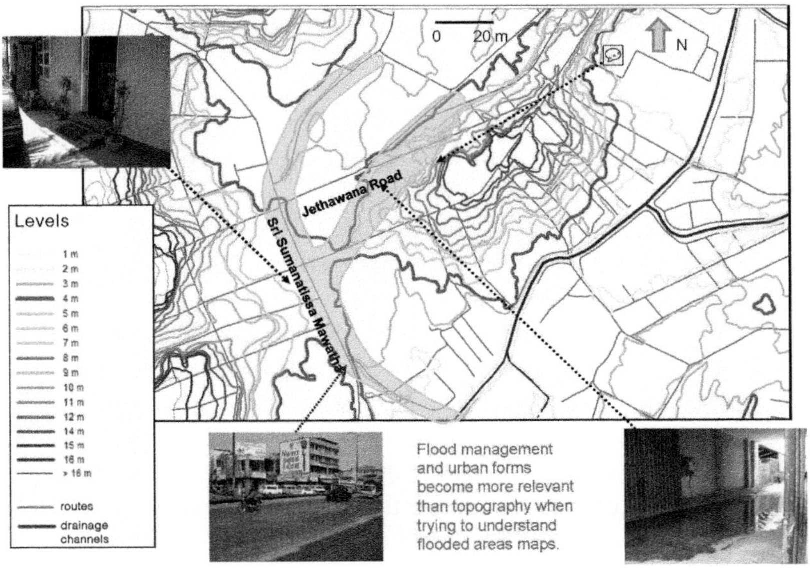

Figure 8.3 : *Le fonctionnement partiel des réseaux d'assainissement contribue à expliquer la répartition des inondations à Colombo.*

Figure 8.3. *The partial operation of sanitation helps to explain the distribution of floods in Colombo*

The limitations that these tools face once again identify their inability to reconsider or even call into question the underlying factors that cause disasters over time. To their detractors, the tools of knowledge management systems are instruments of power, which do not call into question enough the fundamental choices that shape human societies. This brings us back to the need to tackle disasters not in terms of categories, depending on the types of hazard, but rather to evaluate them as components of human settlements, as part of people's relationships with their respective environments. There is also the question of how to account for the limits of disaster prevention policies, while not ignoring their contributions at the same time, during the local decision-making support process. Finally, we describe these limitations, building on the work of Pahl-Wostl [PAH 09]: they explain in part the different types of obstacles currently facing disaster prevention policies, both social and epistemological (section 8.3).

8.1. The knowledge management systems' contributions to disaster prevention policies

The very principle of a knowledge management system is designed to strengthen links between the academic and management sides of disaster risk reduction policies, hoping to improve them. A system emphasizes the relationship between the elements that identify the system and that make up its structure. Knowledge is more related to the academic world than it is to the management perspective. The origin of risk, in connection with the maritime trade, or even gambling, implies the relationship between these two sides (see Chapter 1). The tension between these two aspects of risk is not new, and yet remains problematic. Knowledge management systems claim to provide a new answer to this already very old problem. Gall *et al.* [GAL 15], who are taking stock of international efforts to reduce segmentation between the management and academic sides of disaster prevention, explicitly mention the need to incorporate more science in management and later again in the implementation of policies ("Desired integration of science into practice and policy making").

Other international initiatives are clearly moving in the same direction. In 2013, the first session of a network of experts, professionally involved in the management of information and knowledge dedicated to reducing disaster risk, met in Geneva. This network is formally known as IKM4DRR: Information and Knowledge Management For Disaster Risk Reduction[1]. The first session comprised 60 participants representing 31 countries, some, including one of the authors of this book, on behalf of the academic world. Their purpose was to come to an understanding on what actually constitutes the management of information and knowledge, and the goals that these dedicated systems can achieve. For UNISDR [DUN 14], the main objective is to improve the use of knowledge in the decision-making process by promoting co-production of knowledge "Improving knowledge use in decision-making by fostering the co-production of knowledge" [IKM 13].

Again, this is not new: similar experiments were carried out during the 1970s and applied at least in North America. They were inspired by an understanding of disasters from the systemic and socio-ecological approaches. Tellingly, Gunderson and Holling grant them a prominent place in their treatise published in 2002 [BER 02]. These works were mainly focused on the

1 http://www.preventionweb.net/english/professional/networks/public/ikm4drr/, visited 17 January 2016.

dual objective: on the one hand, to promote greater report sharing among the largest number of actors possible, who are involved in the prevention of environmental disasters. And, on the other hand, it was also tasked to find solutions that collated the greatest possible consensus, through arbitrage guided by decision-making support tools. However, the question of decision-making support for integrated knowledge management is actually much older still. We can concede that it concerns "the formal procedures of public decisions and their underlying normative assumptions – often hidden" [SEN 09, p. 126]. Following this author, we can ascribe "social choice theory" and the work of, among others, Condorcet. We can imagine how the problem posed by policy limitations, surrounding disaster prevention, is as old as it is formidable to understand, and to manage. It is all well and good to bring forward the "underlying normative assumptions" in decisions that mention Sen. The solution that passes through the knowledge management systems is as useful as it is difficult to achieve. It also helps justify Chapters 6 and 7, which focus attention on how tools and policies are conditioned and framed.

Nevertheless, what then is a knowledge management system, practically, and how can it contribute to this age old, persistently difficult problem?

In reality, a knowledge management system comes from the integration of several existing databases, encouraging information sharing, and the creation of knowledge. Recognizing the limits on information and knowledge, sometimes associated with wisdom in the literature in English [WEI 15, p. 109], requires going back to the original databases, adapting them and transforming them. Therefore, "wisdom is evaluated and reflected understanding, or integrated and applied knowledge". It is indicative that we also find Susan Cutter *et al.* [CUT 13] who is very engaged in IRDR (Integrated Research on Disaster Risk) among the proponents of knowledge management systems. IRDR was started in 2008 in reaction to the extreme fragmentation of databases between multiple institutions involved in disaster prevention, across all levels and at all scales. According to Cutter *et al.* [CUT 13]: "Government agencies, private companies and other organizations may collect and manage data related to their own areas of interest using their own standards and procedures, without significant collaboration with other groups. This results in gaps in the data, inconsistent overlaps, and biases that ultimately affect the quality of research conducted and the policies made based on this data". Therefore, knowledge management systems hope to provide an answer to the increasing segmentation amid multiple institutions and stakeholders, including and above all for the benefit of local players engaged in the prevention of disasters. This segmentation contributes to preparing disasters, through the fragmented nature of the different data and information, currently being managed separately by institutions and actors involved in prevention. As it stands, this makes it particularly difficult to have an integrated assessment, even for a particular type of risk; further distancing the possibility of an overall strategic vision, however limited.

One of these knowledge management systems, in France, has been represented at the national level through the "National Observatory of Natural Hazards" (ONRN), since 2012. Managed by the "*mission des risques naturels*", this knowledge management system incorporates several major databases that has been previously separated. In particular, it integrates databases concerning damages recorded by insurance companies in France. The ONRN has helped to provide, in the form of diagnostics, a preview of the results and limitations related to the implementation of a major disaster prevention policy, launched in 1982. The scale of assessment is done not only at the national level, but also at the local level. The French municipality, as a local data cell, allows us to crosscheck data on the number of disasters officially recognized by the French government since 1982, whether or not management tools were in place at the time, likewise risk prevention plans, and the more recent addition, the claims data from insurance companies [NUS 15]. Risk Prevention Plan maps are generally less than 1/10,000th scale to represent the land registry. The ONRN thus provides a stronger, shared database aiding the presentation of an assessment on elements that contribute to the preparation and mitigation of disasters. As such, this knowledge management system can also be used to assess the situation for municipalities with regard to disaster risk reduction policies and their outcomes within the national framework.

Furthermore, knowledge management systems are used as support materials for the decision-making process: they may be able to show the limits of policies and instigate research on how to integrate them into the decision-making process at the local level. We find many examples of feedback on the use of knowledge management systems, for example, in the USA [REN 13, pp. 140–162] or South Africa [REN 13, pp. 164–189). In all cases, these tools are presented as particularly useful, especially when the limits of research and disaster prevention policies are recognized: "planning for hazard mitigation and adaptation can be challenging when there is much uncertainty and disagreement about the best management practices to minimize risks" [REN 13, p. 155]. In addition, knowledge management systems multiply themselves.

The ONRN also allows for building knowledge by the cross-referencing of information from the database with those from local players. This use, crucial to ONRN, is trickier. It is currently underway, and being led by the French Association for Natural Disaster Reduction. For example, we can draw attention to the holes that exist in the national coverage of risk prevention plans, cross-referencing this coverage with the amount of damages that should be reimbursed by insurance once the French State issues a ministerial decree declaring a state of disaster. This situation leads to a greater interest in local challenges in the adoption of this policy. We know, for example, that the catastrophe in Nîmes in 1988 did not allow this town to have a prevention plan before 2012. Such assessments therefore help promote discussions among local stakeholders interested in the limits of disaster prevention policies, and how to adapt them to local constraints, and which have been seen to be very strong. Here, we find the conditionings discussed in section 7.2. The knowledge management system is clearly a powerful tool, in conjunction with its two main contributors: the French State and mutual insurance companies in France. However, it can also contribute to disaster prevention in drawing attention to the parts of the national territory, which unwittingly prepare disasters the most.

We can speculate that the knowledge management systems primarily derive their official justification from the need to integrate as much as possible the databases and the information that comes from the proliferation of institutions and actors involved in the prevention of disasters. These tools are necessarily useful, because we can also demonstrate that institutional segmentation promotes damage and disasters [GUE 14b].

On the other hand, the use of knowledge management systems to help local decision-making, integrating information and knowledge from local actors, still leaves much to be desired; although possible, it is much more difficult to achieve. It can also pose many problems of acceptance. Indeed, it increasingly draws attention to the necessarily limited nature of research as much as of policies aimed at disaster prevention. For example, it is not because a property owner takes into account the requirements RPP (risk prevention plan), as seen in Figure 8.1, that he can hope to avoid all future damage. He remains in the blue zone, which is a flood risk area. It is therefore essential to draw attention to the dual necessity and limited measures that are prescribed by the RPP. This can be simply deduced from the feedback and damage recorded by the monitoring of insurance companies. After the floods in Arles in 2003, and following the work of Domenichini [DOM 08], there appears to be as fairly expected, fundamentally, no significant and usable statistical relationship between the height of the water and the damage as registered by the insurance companies (Figure 2.4). This reflects the plurifactoriality, and complexity, necessary for the interpretation of the damage. Architectural adaptations are made for the best possible outcome: with the aim of reducing damage during a future event, all things considered equal (that is to say, by reducing plurifactoriality). It is assumed that the owner of the second building in Figure 8.1 expected, *a priori*, higher damage from a future flood, while keeping in mind that this will not necessarily be the case. Above all, we cannot expect the RPP (risk prevention plan) requirements to completely eliminate future damages, inasmuch as we cannot expect a dike to eliminate the risk of flood. It is essential for this discussion to be held with local stakeholders. Another major question is asked by the poorest households, for whom the issue of adaptation and its limits may appear secondary, if not even improper, compared with other types of damage that are more immediately present.

In so doing, the limits of knowledge management systems are identified and even denounced. This is certainly the tool that allows fighting against one of the elements that acts in favor of disasters, namely institutional segmentation as well as the fragmentation of information, which is indispensable for knowledge and risk management. However, it is also a tool that, though certainly not alone, fails to reconsider the key drivers associated with the social construction of disasters. Such limits are publicized by the results of research programs, they invite us to address the next point.

8.2. Resistance to prevention policies: poverty and power relations, lifelong obstacles

An example of a brake on the use of knowledge management systems can be seen from the feedback of the COLEM (Colombo environmental management) program, in which one of the authors participated between 2005 and 2009 [CHE 08]. The program aimed to further prevent the risk of disasters concerning the urban agglomeration of Colombo, Sri Lanka. As usual, the principal contribution consisted of a risk assessment, and it was conducted in close partnership with the municipality. This analysis justified the establishment and use of a geographic information system (GIS), located in the municipality of Colombo. This was to allow the monitoring of the different measures proposed, in the event that they are recognized by policy makers. Indeed, the results of the program were presented to representatives of the Colombo Municipal Council. GIS also aimed to follow the settlement developments, to serve as a decision-making support tool regarding the management of civil security.

Involved from the European side were a collaboration from a Parisian university and one of the authors of this book, the involvement of Belgian hydraulic engineers, and a specialist design office engineer for a risk management consulting firm. The engineers of the technical services of the Municipality of Colombo worked in collaboration with the European team. The direction of the program applied was therefore clearly to primarily help the municipality to monitor and to assess disaster prevention policies. The program also represented an opportunity for feedback, especially as it followed on from the 2004 tsunami disaster, strongly resonating with the Indian Ocean community, even though the damage to the coastal part of the Greater Colombo remained limited. The program focused primarily on prevention policies regarding more frequent damages, particularly related to flooding or health risks.

One of the major conclusions from the assessment of the program interconnected some of the earlier work carried out in the framework of international cooperation policies. They were published by organizations such as the German GTZ – *Deutsche Gesellschaft für Technische Zusammenarbeit*; (German Agency for Technical Cooperation) (2005), or the World Bank [WOR 93]. This work from previous assessment, related to cooperation programs, has also been published directly by the institutions responsible for risk management in Colombo, municipal or not. Such was the case for the Sri Lanka Land Reclamation and Development Corporation[2] [SLL 93] or the Ministry of Urban Development, Housing and Building [MUD 00] in connection with Japanese cooperation. All these works draw attention to the tendency of backfilling of wetlands in the coastal plains, which concerns the greater part of Colombo, for the purpose of urbanization. This trend also includes the drainage channels, whose banks are frequently used as a dumping excuse to further build the embankments, but which mainly comprised solid household waste. These same works therefore also seek to attract as much attention to the need to conserve the largest wetlands, lakes and canals, as a means by which to prevent flood risk. The intended use of these channels, lakes and wetlands is contradictory.

The contradiction is reflected in public policies. Even simply as registered in the name and objectives pursued by certain public institutions in charge of both the pilot urbanization and flood prevention, as with the SLLRDC. The official website of the supra-municipal institution, which depends on the Colombo district, shows this contradiction well. The objective to be achieved by this institution is presented as follows: "To be recognized as an esteemed and dynamic engineering organization for the creation and promotion of the development of low lying lands with efficient and effective drainage systems to ensure a flood free habitat"[3]. In short, the SLLRDC is not only responsible for the backfill of wetlands to support the formal land needs of the municipality and metropolitan area, but also responsible for maintaining the main drainage system.

Moreover, poor people tend to settle on the public embankments of lakes, canals and rivers. They use solid waste to gain ground at the expense of wetlands and canals. Poor people are very involved in the process of urbanization which tends to favor the recognition of the damage associated with flooding. Floods are classically expected in a country visited by boreal monsoon.

2 http://www.landreclamation.lk/web/index.php, visited online the 17 January 2016.

3 SLLRDC Website, "About us: vision and mission", http://www.landreclamation.lk/web/index.php?option=com_content&view=article&id=48&Itemid=77&lang=en, visited the 22 January 2016.

Local institutional partners of the COLEM program assume that the reduction of wetlands alongside the channel flow capacity of canals is the principle contributor to the increase in the water level in the main zones affected by the floods. It was not possible to find a measure, let alone a quantified study on this development, which is supposed to promote more damage, and even prepare for disasters. Even the mapping for flooded areas remains sketchy, and accurate information remains scarce, probably due to the politically sensitive nature of preventing this type of damage. The mapping available despite the partnership, which has required months of discussion, only covers the years for which the damages associated with flooding were higher, as in 1992 (Figure 8.2). However, surveys by the engineers in charge of flood risks and health risks in the services that depend on the Colombo Municipal Council (CMC), or those associated with SLLRDC, invariably bring the same assessment. An official letter from two engineers each of the CMC and SLLRDC, and preserved in the municipal archives, though unclassified, carries this assessment which can be considered well established. This letter, in 2000, stated that "the system of storm water channels is being rehabilitated and it is essential that it is serviced regularly. The deposit of solid waste in the canals by local residents has hindered the regular maintenance program. Unless this is stopped in time, the flooding of certain areas in Colombo and its suburbs is inevitable".

Indeed, field work amply confirms the trend officially recognizing the reduction in flow capacity from lakes and canals. It is rare to find channels or lake embankments that are not associated with landfills, and the existence of illegal settlements as well. In Colombo, the latter are also called, revealingly, Under Serviced Settlements (USS). The term refers to the fact that they are unserved by either the water supply networks, or the sewers, and let alone household waste collection services.

However, the fieldwork also shows concerns about the localization of certain areas that were officially flooded in 1992, although not necessarily being in the lowest points topographically (Figure 8.3). A locally conducted survey confirms that the sectors officially recorded on the map are flooded every year. This is consistent with the preventative measures adopted by certain households, such as the establishment of thresholds and cofferdams at the entrance of buildings. However, the field survey confirms that major roads, especially, play an important role during the floods as they channel a portion of the rainfall collected on the ground. Furthermore, the partial operation of the rainwater discharge channels accompanying major highways increases the concentration of water on the latter (Figure 8.3), and strongly contributes to flooding. Illegal settlements and the urban poor cannot themselves explain, from the deposits of household waste, the observed floods.

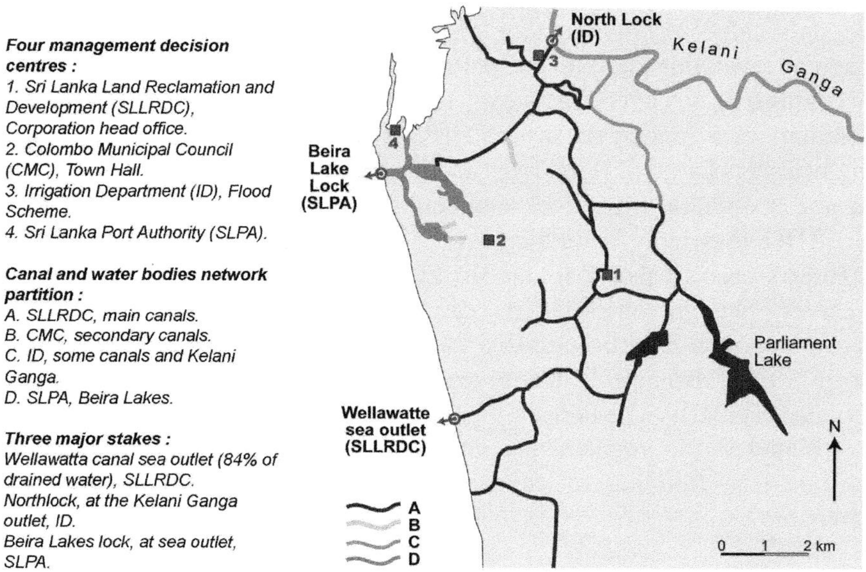

Figure 8.4. *The distribution of the institutional managers involved in the prevention of disasters associated with floods in Colombo. In Pigeon [PIG 08, p. 221]*

If local institutions agree on the official assessment, as shared by international cooperation agencies [MOU 08], they diverge on what action to take. The fields of expertise of these institutions also vary, and they may conflict with each other. The researchers noted fairly early on: working in conjunction with the CMC, which manages the intermediate drainage channels, opens access to the SLLRDC, which manages the main channels, but closes the doors to other institutions involved in the management of flood risks. This was particularly the case at the time of investigations by the UDA ("Urban development authority"), the SLPA ("Sri Lanka port authority") and even the ID ("Irrigation department"). The SLPA is regularly accused by the CMC and SLLRDC of refusing to open the locks that control the water level in the Beira Lakes. These lakes are also associated with the port of Colombo and depend on the SLPA (Figure 8.4). The water level is supposed to be 1.8 meter above the average level of the Indian Ocean. To facilitate the management of drainage, SLRRDC requests that this level of water be reduced, without success at the time of this study. For its part, the "Irrigation department" is a legacy of the paddy fields management on the coastal plain of Colombo, which has virtually disappeared in the city. However, ID still controls a canal lock on the San Sebastian canal, which opens on the Kelani Ganga. In addition, both institutions, the SLPA and ID, argue that lowering lake levels and channels increases the destabilization of lake and canal embankments. Such an argument can be defended because the embankments are almost always reinforced by protective structures such as gabions or sheet piles (steel walls reinforcing the lake or Wellawatta canal embankments, as in Figure 8.6).

Figure 8.4 shows that the institutional fragmentation of information is not a vain word in Colombo, and does not facilitate the integrated management of disaster risks associated with flooding. It is fueled by the tensions and rivalries between these different institutions. In these circumstances, there is no doubt that establishing a knowledge management system would basically make sense, and hopefully fight against such fragmentation. However, without strong political support, this approach could not be any more successful than the one encountered by the GIS, located at CMC. Between 2005 and 2009, GIS was not available to all departments working for CMC. The unit for health risk complained openly, and developed its own database (Figure 8.5). Now, especially in Colombo, where cholera is endemic, and where other diseases are clearly linked to the quality and management of water, managing flood risks requires crosschecking information with that concerning health risks. Therefore, we can see how a system of knowledge management tool can contribute more to disaster prevention, provided that it receives the political will/backing to give it more meaning. This backing will be embodied in facilities, in institutions that actually work together in co-constructing shared knowledge on disaster prevention and including the effective participation by local populations. This clearly was not the case during the COLEM program, despite the official support of the municipality.

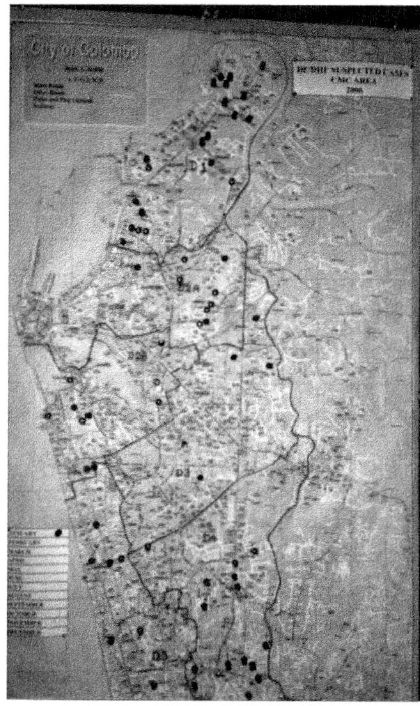

Figure 8.5. *Map showing the distribution of dengue cases in Colombo in January–February 2006 (in Pigeon [PIG 08, p. 226] with permission from Dr. Pradeep Kariyawasam reproduction, CMC). This is a wall map on which deferred dengue cases were confirmed in January 2006 (black pins) and in early February 2006 (blue pins). For a color version of this figure, see www.iste.co.uk/rebotier/disaster.zip*

The other major obstacle to explain the discrepancy between what is expected of prevention policies and what is recorded *a posteriori* is clearly associated with poverty. Even in neighborhoods undergoing gentrification, that is to say, which are marked by the trend of relocating the poorest in the face of increased property values and well-off populations, we can always observe even briefly the maintenance of illegal discharges. An example is provided by the embankments of the Wellawatta canal. The latter is the main drain of the southern part of the coastal plain of Colombo, and the outlet channel is not controlled by a canal lock. This canal is fully managed by the SLLRDC. Upstream of the canal is the national parliament of Sri Lanka, in the town of Sri Jayewardenepura Kotte. In conjunction with the CMC, the manager of this canal, the SLLRDC, conducted a dredging policy and monitoring of the embankments. It was also accompanied by the relocation of the poor previously occupying the public embankments of the canal, as well as a nearby factory for spinning and weaving. The latter ceased operations, and the site is now occupied by the town planning operation “Havelock city”.

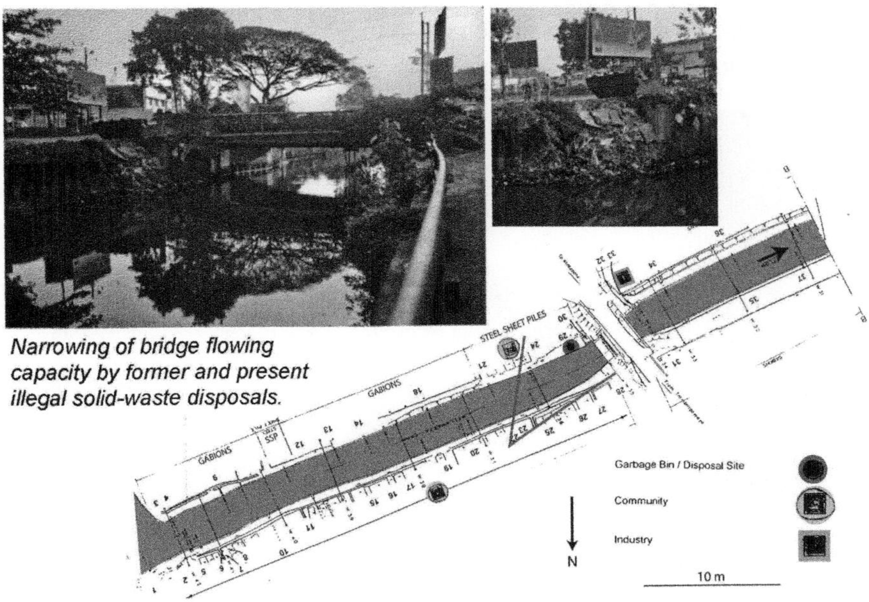

Figure 8.6. *The boundaries for the policies of the SLLRDC and CMC in the Havelock neighborhood on the banks of the Wellawatta canal (in Pigeon [PIG 08, pp. 217–218]). The dumpster for collecting household waste, carried on the official plan, is not sufficient to rule out any illegal deposit, and this is the case here on the upstream abutment on the left bank of Wellawatta canal. The canal’s flow capacity is reduced*

However, the disappearance of these poor people illegally occupying the channel’s public embankments did not signal the end of illegal deposits of household waste. A very aggravating circumstance was that these deposits continued on the left embankment abutment by the Havelock bridge (Figure 8.6). As shown in the map found in the archives of the SLLRDC, not yet archived, this site is subject to special monitoring, and a dumpster was even installed there. Fieldwork confirms that the measures taken do not come close to achieving the objectives. It is clear that the illegal deposition further reduces the flow capacity of the Wellawatta canal. It accentuates the narrowing of the canal and therefore its flow capacity, both visible in plain sight, as shown by the photograph. Even in an area marked by gentrification, the limits for garbage collection in Colombo seem to signal the continued presence of households with limited resources. A simple causal relationship between poverty and deposits of household waste clearly does not exist. Yet, the trend toward reducing canal flow capacity by deposits of household waste is obviously even more pronounced where the poorest people are, as on the embankments of the San Sebastian channel up to its confluence with the Kelani Ganga (Figure 8.7).

Figure 8.7. *Tendency for reduced flow capacity of North San Sebastian canal, at the confluence with the Kelani Ganga. Colombo. P. Pigeon, February 2006*

Poor people do represent a major element for understanding why the prevention policies that were led by CMC and SLLRDC are limited. Moreover, these public institutions defer the liability of the limits of the measures taken by implementation of prevention policies on these poor people – especially if they are Tamil, and even more belonging to a minority, as is the case here [MAD 13]. Institutional managers seem to follow to the letter the "first differentiations of the theory of liability" as proposed by Jonas [JON 79, p. 179]: "the responsibility condition is the causal power. The actor must answer for his act: he is held responsible for its consequences and if necessary we make him bear responsibility". However, Jonas also reflects on the links between responsibility and the "obligation of power". In addition, what is the power of these poor people? We then find a question of reading what could be that of Sen, from the "Capabilities"

[SEN 09, p. 313]: "real poverty (in terms of deprivation of capabilities) is much more profound than that which we can deduct from income figures". In these circumstances, responsibility is a function of capabilities, such as the consideration by Sen [SEN 09, p. 45]: "the freedom of choice allows us to decide what we should do, but it comes with the responsibility of what we do – in so far as it stems from a truly chosen action. Since a capability is the power to do something, responsibility emanating from that power is part of the problem of capability". The more the capabilities are reduced, as a marker of poverty beyond income criteria alone, emphasizing the reduction in choice, the more it is simplistic and ethically questionable to blame limitations ... on the poor. Fundamentally, the simple and sufficient causal relationship between poverty and the limits of prevention policies is untenable. Quite simply, it disregards the plurifactoriality, which presides over the creation of risks and their management.

Under these conditions, how are we able to exploit the potential of a knowledge management system involving these poor people, who live in precarious conditions every day, and with very limited choice of capacities? How should data and information be shared in order to build a co-constructed knowledge on flood-related disaster prevention policy whereas there are no statistics on damage affecting the poorest populations? The municipality of Colombo has little information available on the health risks of the poor. The existing view of policy makers and management is certainly heading for dengue fever, which discriminates less between people based on their living standards than flooding does. Among the technical services of the Municipality of Colombo, it is said that *Aedes aegypti* has the bad taste to sting the rich as much as the poor. As for the rest, the embassy district south of the city and near the Wellawatta canal is subject to special supervision, financed in part by the Rotary Club. Cholera is considered less of a priority. However, it concerns the poorest first. Therefore, we return to the problems with the inclusion of extensive risks as outlined in the classification of UNISDR, discussed here in the second part of this book, and that of their reductionist databases. As the knowledge management systems are mainly based on the integration of existing databases, we once more know their limits.

We better understand why some of the literature considers the knowledge management system as a tool, with a very critical eye. Indeed, despite a potential integration of information and dispersed knowledge, nothing indicates that the KMS will not stumble on the same order of political or institutional deadlock. To try to not repeat the recurring pitfalls through yet another scheme or alternative approach, the limits of KMS, as with the limits of disaster prevention policies, need to advocate for better political representation, in order to help to better understand them in a more integrated way.

8.3. Evolution or revolution? The dilemma of Pahl-Wostl

The desire to have the most coherent vision possible for the limitations of disaster prevention policies, while recognizing their contribution to limiting losses, at least in relative terms, requires new advocacy efforts. This search for clarity is desired among many other types of policies, and is notably in agreement with the work of Revault D'Allonnes [REV 02, p. 149]. For this philosopher, "historical time is in fact a political time" because we have to face the necessarily limited nature of political actions. Feedback, once again, is essential here because it proves the existence of these limitations. Of course, this requires being able to take on board feedback, and also have an interest in doing so. The literature includes at least one solution to the problem of representation.

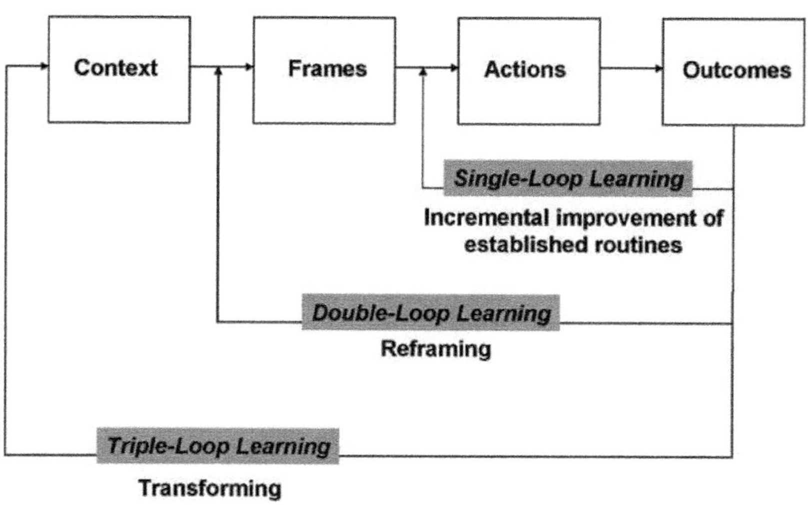

Figure 8.8. *The conceptual model of Pahl-Wostl [PAH 09, p. 359]*

Based on several previous works (including not only Argyris and Schon [ARG 78], but also Berkes and Folke [BER 02], who belong to a group of researchers under Resilience Alliance) Pahl-Wostl [PAH 09] proposes a scheme of three feedback loops (Figure 8.8). This type of scheme, which Pahl-Wostl called a conceptual model, corresponds to our needs. Pahl-Wostl participated in the collective work of Rockström *et al.* [ROC 14]. These authors worked markedly, and also particularly, on governance issues associated with disaster prevention and related to the watersheds of rivers. Incidentally, this also explains the proximity of thought with researchers who reflect on the prevention of disasters associated with water courses by referring to socio-ecological systems, like Cumming [CUM 11]. This type of pattern was also reflected in a report of the European funded program Cap-Haz net [KUH 10]. The diagram in Figure 8.8 is not isolated in the literature on disaster prevention, whether the researchers claim themselves as belonging to the Resilience Alliance group or not, or whether they are working on the prevention of disasters associated with water courses or not. We guess that it can perhaps be read in a differentiated way, something which Pahl-Wostl herself admits.

The illustration depicts three types of feedback loops. They represent several ways to consider and reconsider policies and the results that their applications deliver. The Pahl-Wostl model suggests an evolution, which inherently involves a quest for improvement. The recognition of the limitations faced by these policies requires consideration, and can be seen to be part of a learning process. As Pahl-Wostl [PAH 09, p. 358] states, "Learning is assumed to be an exploratory, stepwise search process where actors experiment with innovation until they meet new constraints and new boundaries". The interpretation is thus systemic, thanks to feedback loops, as well as being dynamic. The representation is more consistent with the idea of feedback, or with post-disaster reconstruction policies. The latter necessarily reconsider the settlements that existed prior to the disasters. The diagram that Pahl-Wostl shows simply formalizes the difficulties so as to push the limitations that disaster prevention policies meet. This reconsideration of the limitations of policies allows for an explanation, at least in part, of the results that disaster prevention can achieve. Indeed, we can dispute that the more we know, the more we manage, and ultimately the more we lose.

Let us examine how this "conceptual model", according to Pahl-Wostl, can represent these elements, and justify why, by revisiting field examples.

The first feedback loop (Figure 8.8) is an iterative, incremental process, to gradually improve the results obtained by a disaster prevention policy, but without altering the fundamentals: "Single-loop learning refers to a refinement of measures to improve performance without changing guiding assumptions and calling into questions established routines" [PAH 09, p. 359]. Although measures taken previously appear limited, it may in fact correspond to what is found when several, identical, measures are taken successively for a solution of a collectively identified problem.

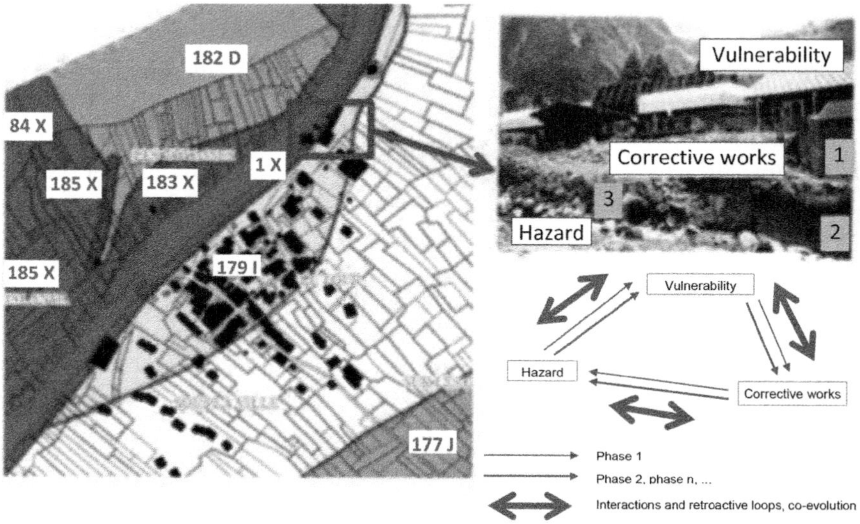

Figure 8.9. *An example of polychronic corrective work. Chamonix Le Tour, 2002. Pigeon [PIG 14] https://espacepolitique.revues.org/docannexe/image/3256/img-2.png*

COMMENT.– *The photograph, taken in 2002 and located on the Chamonix flood risk map, approved May 17 2002, highlights at least three generations of dikes which appear to separate the torrential river Arve and the buildings of the village of Le Tour. However, consultation of the Chamonix Municipal Council archives, as well as those of the restoration of the mountain land* (service de restauration des terrains en montagne) *archives service, in Annecy, allows one to verify how the dikes, the river, and the buildings interact. The archives allow us to date the corrective works (dikes), as well as the urbanization phases. The archives also mention the trend towards the channelization of the Arve minor bed, which promoted the three phases of diking, visible on the ground, and illustrated here in the photograph (Pigeon, 2002). The partial justification of these corrective works was for the cable car development (as well as the increased security). But the dikes also unwillingly contributed to an increased riverbed incision of the channelized torrential river. In turn, this loosening of dikes by incision is proved by the three phases of works identified in the field, and which the archives mention.*

This exemplifies how a seemingly static element such as corrective works (dikes), a torrential river and buildings historically interact, and how the political decisions taken did not reconsider the "established routines", in spite of various generations of damages experienced. This is in line with the first feed-back loop the Pahl-Wostl model displays.

An example of this can be provided by the experiences of corrective works, such as dikes. Several generations of corrective works are occasionally visible during field surveys (Figure 8.9). The proliferation of corrective works visible on the ground is documented and dated by the archives of the institutions involved in the implementation of the approved policy. These multiple corrective works illustrate both the decisions and limitations of policy. In Figure 8.9, the river Arve has been channelized, and this channeling of the torrential river has favored the incision of the Arve riverbed. As a consequence, the diked embankments have become destabilized and threaten to break up. Of course, it was not possible to leave the situation as is, since the diked embankments are here part of the corrective and protection works. The diked embankments protect nearby buildings, which include the Balme-Charamillon cable car. This landmark provides access to one of the few ski areas of Chamonix accessible to beginners. Moreover, dikes that protect the parking lots, essential for access to the ski area, have been gradually straightened, thereby increasing the flow into the Arve. On several occasions, the reaction by the Chamonix City Council, as with the service of the Haute-Savoie in charge of risk prevention ("*service de restauration des terrains de montagne de la Haute-Savoie*"), was to strengthen the embankments threatened by these new developments. The political response to disaster prevention of the Arve flooding therefore can be seen to return to the same type of measure: the correction of the torrential river by works such as dikes, which also act as protection works. Reconstruction is the most similar, though it is not identical.

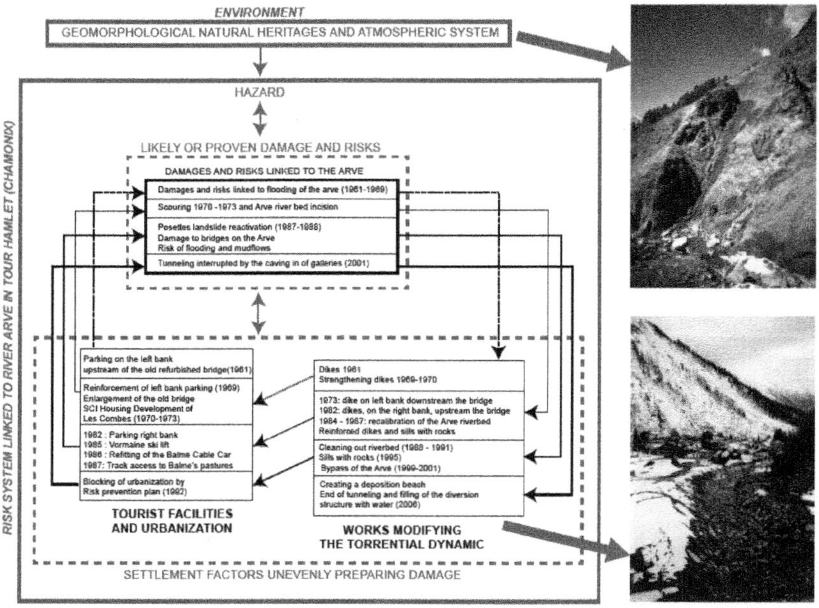

Figure 8.10. *Representation of positive feedback loops in Le Tour, Chamonix [PIG 02]*

We can further explain why this type of response has been given. The different phases of corrective works are typically associated with urbanization phases, which are found in both field surveys and research in the archives (Figure 8.10 gives an example of this type of positive coevolution for the site at the hamlet of Le Tour, located at Chamonix). Here, there are several important interest and economic factors in play, which are associated with the disaster prevention process, particularly the corrective works. There are the landowners who are behind the diked embankments, the real estate developers, and building companies and public works that build dikes, roads, and parking lots, not forgetting the design agencies which license them. Moreover, the existence of the corrective work allows the Chamonix municipality to hope for the representative of the French government to agree to shrink or lower the requirements of risk zoning prevention plans display. Behind an embankment, it seems logical to have no red zone, which is not to be built on, but a blue area. Remember that the blue area remains buildable, subject to consideration for the risk that is officially recognized, which is accordingly reduced by the existence of the diked embankment (or any other type of corrective work).

Of course, it is understood that, eventually, the question needs to be raised as to whether it is appropriate to identically reproduce such a prevention policy, given the motivation behind the corrective works involved. Indeed, as we have seen, prevention strategies stemming from corrective works can unwillingly contribute to preparing disasters. This is particularly true in the cases where the limitations of corrective works are not taken into account by land use policies. Here, we find another dimension of corrective works, as with dike risk which has been presented above (see section 4.2). Dike risk recognition may be pushed by other major players of prevention, like the French state or even insurance companies. Corrective works may also be challenged by environmental protection associations, such as associations of fishermen. Moreover, the cost of these corrective works is also high. This is particularly significant if they are to be financed by the budgets of local authorities, be it the town hall or inter-municipality.

Then comes a possible justification of the second type of feedback loop, in agreement, at least partially, with the model taken over and modified by Pahl-Wostl [PAH 09, Figure 8.8]. This time, the reasons behind the previously conducted policies are reconsidered, as is the paradigm, or framework of thought, in which they occurred, but which were not fundamentally transformed, in spite of the limitations and damages experienced. According to Pahl-Wostl [PAH 09, p. 359]: "Double-loop learning refers to a revisiting of assumptions (e.g. about cause–effect relationships) within a value-normative framework". The questioning of disaster prevention policies via corrective works admits that the solutions cannot be purely local, nor that they cannot be based on simple and linear causal relationships. The diked embankment does not eliminate damages. In this hazard-centered reading, damages are causally attributed to flooding and to the torrential river, which represent the hazard. The systematic approach serving as the base for the prevention policies, through corrective works, is in itself prolonged by the systemic approach. It replaces the corrective and protection works, such as dikes, within a collection of other settlement components with which they interact and, similarly, co-evolve with (Figures 8.9 and 8.10). The damage can neither be explained nor be managed only by examining the hazard, even if we cannot explain or manage it excluding the hazard. The systematic approach is not contradictory, nor a total break with the systemic approach. Similarly, the reconsidered policies do not become completely isolated from those that preceded them.

Let us illustrate this with reference to tangible examples and types of policies widely recognized today. Disputes among prevention policies that focus on corrective works have become so numerous that they have led to a shift in policies. In the field of disaster prevention linked to flooding, these policies, now sometimes call for the availability of open areas for watercourses (Pigeon in Warner *et al.* [WAR 13]). The idea was to dismantle some of the corrective works, such as dikes, which were previously seen as key for defense. Depending on the issues at stake, some of the dikes could be left untended, allowing the same river to gradually destroy them. As a result, the river recovers some of its major flood plain, previously occupied by the diked embankments themselves. This type of policy hopes to reduce the intensity of future damage to towns downstream, by facilitating the storage of a portion of the water in the alluvium of the bed of the river, thereby reducing the speed and height reached by the flood wave. The aim is to prevent even more disasters.

The paradigm is adjusted, because there is no fundamental reconsideration of the potential role that corrective works play, or even of the decision-making process, which is still very top-down despite the effort of renewing governance. This paradigm adjustment does not apply solely to the prevention of disasters associated with flooding. Another equivalent is represented by the strategic withdrawal coastal policies may implement. We also note that this paradigm adjustment of policies admits not only their inability to eliminate damage, but also their ability to reduce the intensity of future damage, or in part, to prevent disasters either.

The tools that are knowledge management systems are favored by this transformation of disaster prevention policies. There is a need to find tools which integrate multiple databases from different institutions, working at multiple scales, and whose institutional interests do not necessarily converge. In the reading that Pahl-Wostl gives with her "conceptual model", knowledge management systems are attached to the third type of feedback loop (Figure 8.8). It refers to the following criteria: "Formalized participation of actors at different levels; Established practices of knowledge exchange across levels; More polycentric structures and balance between bottom-up and top-down approaches" [PAH 09, p. 360]. It may then seem tenable to justify the third feedback loop, and learning, from these tools.

However, as we have seen, the tools that are themselves these knowledge management systems are focused primarily on macro institutions, more than they are toward the inclusion of local actors. The literature is unanimous on the very limited remaining space available to these local actors, within the decision-making processes that accompany disaster prevention policies. The post-disaster feedback also increasingly recognizes that it will hardly be possible to expect further reduction in the losses associated with future disasters, without taking into account the manner in which information is shared and integrated among local populations. Such assessments have value whatever be the type of organizational power. Above all, they are not the preserve of State centralized structures, or today decentralized ones, as it is the case in France. Indeed, we discover the same problem for a federal state, like the United States. This is well demonstrated by the thesis of Emeline Comby [COM 15], which compares the policies related to the management of water between the Californian Sacramento and the French Rhone, and the involvement of local communities in policy decision-making remains very low in both cases. The thesis thus echoes with one of the conclusions from the KNOW4DRR program. The program compared thirteen case studies on the management of information and knowledge related to disaster prevention, and from countries with as different a political structure as Vietnam, Germany, Mexico, Austria, France and Italy. In all cases, federal or not, the information and knowledge integration from local people remained under considered, and in general, they were poorly integrated into the disaster prevention policies affecting them.

These findings are consistent with the poor place devoted to “extensive risk”, or “small disasters”, according to the terminology of UNISDR, to be found in existing databases, with the exception of DesInventar, and with the criticism of the Sendai framework for action. For example, according to Zia and Wagner [ZIA 15, p. 197]: “Although the SFDRR includes people centered, bottom-up rhetoric, it lacks explicit means of implementation for governance structures that incorporate people-centered, bottom-up design”. Similarly, “A true test of SFDRR implementation will be if investments and resources are allocated in such a way that local vulnerable communities are empowered to increase community resilience and DRR with or without the support of relevant national or regional entities”.

This brings us back to a fundamental critique given to disaster prevention policies and the tools they can mobilize. The tools that are themselves these knowledge management systems can deliver diagnostics on prevention policies by integrating previous databases, and at least some of the time, can reduce institutional fragmentation. The contribution of this fragmentation when it comes to the production of disaster is amply demonstrated by the feedback from multiple post-disaster experiences. The tools that are KMS, such as the ONRN, are essential to the prevention of disasters. By contrast, these tools do not take into account, as such, other root causes that prepare disasters whatever be the type of the hazard implied. In this understanding of knowledge management systems, the shifts from the hazard-centered paradigm, which symbolizes the distinction between the first and second feedback loops, may seem to justify the third feedback loop. But in the Pahl-Wostl model, "Triple-loop learning refers to a transformation of the structural context and factors that determine the frame of reference" [PAH 09, p. 359]. According to this author, it is a "societal learning" which involves all components of policies, including their major factors. The difference, established by Pahl-Wostl, between social and societal, looks to draw attention to the need for shared learning more collectively as a fundamental driver behind a change in structure that challenges current forms of governance; "The distinction between social and societal is made to emphasize the importance of learning in multi-actor settings and of structural change in the governance regime as a whole" [PAH 09, p. 358].

However, it is doubtful, in the light of available and present feedback, that knowledge management-related tools can achieve this objective effectively. We are most often confronted with a liberal variation of the appreciation for local and alternative sources of knowledge, which decollectivizes and individualizes the understanding, if not responsibilities, by putting the blame on individuals for situations of risk and their management. It is in any case the way some authors understand the newer disaster prevention mechanisms [QUE 15], or similarly, the flagship notions used both for the analysis and action on environmental issues and risks [FEL 14].

Indeed, the hope to prevent more disasters requires a deep reconsideration on the power relations between institutional actors, who dominate political choice, and others, especially at the local level. We find in particular, the question of what may be at the heart of empowerment of the poorest populations; the notions of risk and uncertainty that have little meaning if we consider the poor capacity for choice that characterizes these populations [JOU 06]. They can hardly use the post-disaster reconstruction phases to stop, in a collectively responsible manner, the engine that contributes to the fabric of future disasters, and which ultimately is at the very heart of local settlement systems.

Conclusion

"It is generally well known, that precisely because something is well known, it is not known. This is the most common way to deceive and mislead others as to presuppose something of knowledge is well known, in order that it becomes tolerated as such"

Hegel J.F., *La phénoménologie de l'esprit*, Aubier, Paris, Volume 1, p. 28.

Why a further summary for the theme of disaster risk prevention and what are its contributions?

The justification for this book is born of the participation of two individuals who have authored more than a dozen national and international research programs on the issue, not to mention the dozens of conferences and seminars that accompany them... The daily bread of the researcher on the prevention of disasters today. In the light of these exchanges, at least two problems attracted our attention. Often in discussions, especially when informal, our counterparts admit that the two issues in question are placed in an unsatisfactory manner and thus open to debate. However, these two problems paradoxically continue to support and even to structure both the research and the management that are devoted to disaster prevention.

Two problems

What are they?

The first problem, or paradox, *assumes that one always loses more than one expects, whilst understanding and preventing disasters even more,* as if the policy and academic work were ineffective. This element is justified by referring to a widely cited article, at least 268 times as Google Scholar, the White *et al.* [WHI 01]. Although this first problem looks like a false paradox as we can find logical solutions to it [VID 04, p. 8].

The second problem is related to *the very unsatisfactory nature of definitions, and therefore also the conceptual models concerning the disaster risk prevention*. They are nonetheless commonly used in research, and included in the programs that support national and international institutions. Again, there is much work which already exists on the subject, including that of Birkmann [BIR 06]. However, this type of problem, described as very difficult, is not likely a false paradox in the logical sense: which by definition should resist any attempt by logical comprehension.

The two issues are linked, and even venerated as "well known" within the goal statements of research programs, because the latter assumes that a better understanding would infer better management, and ultimately results in less loss. By contrast, we defend that these two problems, as they are usually postured, and represented, are at least in part, a means by which not to address key issues in the field of disaster prevention. As such, several objectives justify our contribution, and this book.

Five objectives

A fundamental objective is to *specify the components of these two issues, so as to test their acceptability*. The first problem reveals a large ambiguity, which adds to the false paradox: it shows that it is not only a false problem, nor a problem only related to a too rapid reading of the matter. On the one hand, in the first chapter of this book, we audited the assessments which show that more disasters means more prevention policies, and therefore more research, yet without experiencing a reduction of disasters. These assessments are generally transparent, easy to handle, predictable and all too many. On the other hand, the assessments, when looked at closer, become much more nuanced. The effort for a critical reading of databases in Chapter 2 shows that disasters are not always on the rise, no more than is mortality, even when we take into account the share of hidden mortality. Even where losses have increased, there is often more to lose again, not that this should lead us to consider these losses as acceptable. There are also a large number of markets related to the prevention of disasters, which take advantage of this growth in activity. It is difficult to admit that all these efforts would be useless, if we agree with the paradox that the more we know, the more we manage, and, yet, the more we lose... On the contrary, it all suggests that disaster prevention is effective, albeit with multiple imperfections.

Reconsidering the first problem induces us not to understand or manage a disaster in isolation, in an absolute way, but as a component of human settlements dynamics, therefore, in a relative and contextualized way. Disaster is not considered in isolation, as such, but according to how it is evaluated, or through the amount of effort dispensed to prevent it, or in terms of what there is to lose. As a counterpart, the topic of disasters becomes even more difficult to define, understand and quantify from databases that classify, divide, separate. Bringing us to our second problem, that of related definitions and unsatisfactory conceptual models, the two issues are in fact closely related. After the first part of the book, disaster prevention remains paradoxical, because you have to simultaneously understand the possibility to prevent disasters, but without hoping to completely eliminate disaster. In addition, this is why sometimes, while seeking to prevent the disasters, a policy helps produce them...

The second objective therefore seems to ask: how can we grasp that these assessments are as ambiguous and difficult to establish, beyond the obvious ethical and economic issues? How can we explain and act at the same time, while meeting as many limitations for the action as for the explanation? Our main response overlaps that provided by the literature: the disasters are considered and managed as absolute, not relative to a context that gives them meaning. Above all, they are basically designed according to the hazard (explosions, riots, earthquakes, insect infestation, virus, etc.), which categorizes them, and thus they continue to be managed by specifically dedicated research and institutions. The approach is classic and seemingly natural. However, it is also very limited. Greater disaster warning requires us to envisage and manage them as human settlement components, and in their own context. For UNISDR, as seen through the eyes of this so-called radical approach to the issue [WIS 04, LAV 13], disaster prevention is a development issue which deserves to be expanded hereafter, much beyond the idea of disaster as a unique and isolated problem. It is also a problem of environmental management.

Bringing us to the third objective: how can we understand that the understanding and the management of disasters are always addressed in a very compartmentalized and segmented way, and yet acknowledge that the limitations of these approaches are recognized? The answer that we defend here is at least twofold. On the more *academic* level, a problem requires it to go through a phase of systematic analysis. However, does this systematic necessity tend, as highlighted by many philosophers for centuries, toward the myopic? Does it not also tend to freeze what is evolving? The bridging of this gap is very limited, especially in the field that interests us, that of disaster prevention. System analysis and complex analysis can complement and compensate for the limitations of systematic analysis. However, as clearly mentioned, in particular and at many points by Bergson, we remain in the field of analysis. In addition, analysis segments and fixes, no matter what we do. Therefore, analysis is not completely adapted to the understanding and management of disaster prevention, which demand a broader and more dynamic reading.

Aggravating circumstances, this analytical and categorical specialization is found in the *management-oriented* dimension of disaster prevention. As we have verified in this book, an institution specializing in the prevention of a type of disaster does not necessarily work closely with another, even if the type of disaster is the same or related. Disaster prevention is thus read in the light of multiple institutional interests. It is frequently hampered simply by conflicts of interest, or power relations, which certainly prevail over inter-related objectives of the specific disaster prevention, despite all being linked *de facto*.

The *specialization* and the *fragmentation* that mask the connection of multiple objectives among which is included disaster prevention therefore remain a fundamental part of the problem, one probably as old as humanity's ability to think and act. The limitations on action and reflection are more quickly found when taking various scales to understand and manage disasters. It can also involve multiple time-periods, as well as many political participants whose interests, by definition, rarely converge. The price to pay to be able to think and act is necessarily imperfect, which we should seek to acknowledge rather than deny or ignore.

From here, we can find the fourth objective: how can we think and act in spite of limitations, simplifiable and manageable, but never completely eliminated? We present these current solutions for the field of disaster prevention in the form of knowledge management systems. These tools are actually meta databases, seeking to *reduce the limitations of knowledge and prevention policies that come from fragmentation*, both academic and institutional. They help provide new knowledge that is more integrated and less segmented. Despite the considerable methodological limitations that these integration tools face, KMS allows the reconstruction and approximation of the underlying trends that accompany disaster prevention policies over time. This is one of the results provided by the *Observatoire National des Risques Naturels* (or ONRN) in France, for example. However, these tools also face other limitations such as those found in the socio-economic inequalities or power relations in the decision-making process associated with prevention policies. For prevention policies concerning the very poor households, the notion of disaster has no more meaning than risk or even less uncertainty. Ascribing culpability to these populations becomes akin to an ethical scandal.

Thus, if the working perspectives exist to counter the limitations of specialization and fragmentation, it is always at the price of the imperfection, arbitration and/or inequality that are necessarily built in the context of the prevention policy.

The fifth objective can thus be defended: as a specific contribution from the human and social sciences to the question posed by disaster prevention. The human and social sciences' constructivist ambitions are working to understand the plurality (and sometimes even the contradictions) of the meanings, interests and the conditions, which characterize the disaster prevention. The produced knowledge, socially and politically conditioned and contextualized, not only deals with the disasters, but also considers the context, arbitration and various interests through which they operate. This *broader integrated understanding concerns equally the types of human settlements and their production of knowledge on disasters and their prevention*. For it is by making an account of the related conditions surrounding disasters, particularly those that experience dissatisfaction, that new policies can be inspired to not be victim to overspecialization and over-fragmentation. The human and social sciences can thus provide additional focus to the technical perspective that is absolutely necessary, and at the same time irreducibly inadequate, given the previously identified limitations. Both the systematic and technical scope, as with the comprehensive scope and contextualized knowledge, are nevertheless part of the analytic registry. Therefore, neither approaches the reality and complexity of disasters and their prevention enough. However, the specific knowledges of the human and social sciences are entitled to make an original contribution to the theme, while scrupulously denaturalizing the paradoxical data and politicizing them [HUL 09].

Perspectives

Yet, we often oppose the delicacy and complexity of understanding to the urgency to act and its consequence: simplification. How should the contributions of this new summary to disaster prevention in the context of research and future action be considered?

The integration and return of plurality, ambiguity or contradiction is not an admission of failure or impotence. They simply take note of the unattainable character of utopians (that of perfection, zero risk and eliminated uncertainties), and at the same time, they recognize the highly political dimension of disaster prevention. It is meant by this that disaster prevention can only be envisaged, for example, in unequal social relations, or in a social and historical context of habits and unwritten rules [FOR 12]. The questioning of the mechanisms of the hazard, or the impact prediction is then not first and foremost, but rather affiliated.

In the first place, and echoing the quote above that introduces the conclusion of this book, these additions to understanding may give the illusion of making the solutions even more difficult according to the limitations of the disaster prevention. They lead away from the solid perspectives of systematic or technical approaches on which are based the necessary – and comforting! – responses to the disaster in question. Yet if we take note of the limitations of disaster prevention, which can never be completely removed, and the recurring deadlocks that maintain the paradox as discussed in this book, we can see the value both for understanding and for action, as provided by this additional knowledge from the human and social sciences. The improvement of disaster prevention is not only understood in the elimination – utopian – of uncertainties or in the progression of – infinite – knowledge. It is also in the discussion of values and social orders, which govern arbitrations and the priorities among which the policies of disaster prevention take root.

Thus, another arena opens up and gives room for a scientific activity that is less dominated by the analytical approach, simple causalities or technical objectives, and that is rather more open to plurality, more responsive to axiological discussions, and more engaged with the political significance of science. There is no shortage of examples of people opposing the implementation of policies that have been decided unilaterally and without taking into account local conditions, or that claim to eliminate the damage targeted without trying to bring forward their own limitations. The knowledge produced (and the manner in which it exists) is not without any effect on society. The knowledge management systems, for example, can feed into many different scientific disciplines, ranging from the more technical and instrumental, to the more political and axiological. Yet, contributing to the production of additional knowledge on disaster prevention will not happen by itself. It is nothing but one possible way of understanding among others, identified here as the human and social scientific approach, and it will be for everyone to seize the action in a voluntary, rational, manner.

Thus, there is no "solution" to the final or absolute paradox. There are simply alternative insights that lead to other tools, or other forms of knowledge accounting for interests that must be made explicit.

It is at the price of this highly political dimension that tools such as KMS can make a difference with regard to the recurrent paradox, and contribute to the task of better preventing disasters.

It is also through such insights, and thanks to the additional illumination of critical obstacles, hampering the improvement of disaster prevention, under which the social sciences may justify their specific contributions to the thought and action surrounding risks and the environment. Disaster prevention policies are bound to meet limits, but not illuminating and discussing those limits hampers reducing them even more. This is why the paradox of losing more while thinking and managing more should basically be questioned much more than what it is as now.

Bibliography

[ADG 06] ADGER N.W., "Vulnerability", *Global Environmental Change*, vol. 16, no. 3, pp. 268–281, 2006.

[ADG 09] ADGER N.W., DESSAI S., GOULDEN M. *et al.*, "Are there social limits to adaptation to climate change?", *Climatic Change*, vol. 93, nos. 3–4, pp. 335–354, 2009.

[AIT 10] AITKEN M., "Why we still don't understand the social aspects of wind power: a critique of key assumptions within the literature", *Energy Policy*, vol. 38, pp. 1834–1841, 2010.

[ALB 02] ALBOUY F.-X., *Le temps des catastrophes*, Descartes et Cie, Paris, 2002.

[ALC 08] ALCOFORADO M.J., ANDRADE H., "Global warming and the urban heat island", in MARZLUFF J.M. (ed.), *Urban Ecology*, Springer, New York, 2008.

[ALD 06] ALDHUY J., "Modes de connaissances, intérêts de connaître et géographie sociale", in SÉCHET R., VESCHAMBRE V. (eds), *Penser et faire la géographie sociale. Contributions à une épistémologie de la géographie*, Presses Universitaires de Rennes, 2006.

[ALE 13] ALEXANDER D., "Resilience and disaster risk reduction: an etymological journey", *Natural Hazards and Earth System Sciences*, vol. 13, pp. 2707–2716, 2013.

[AND 13] ANDRÉ C., Analyse des dommages liés aux submersions marines et évaluation des coûts induits aux habitations à partir de données d'assurance: perspectives apportées par les tempêtes Johanna (2008) et Xynthia (2010), PhD Thesis, University of Western Brittany, 2013.

[ARG 78] ARGYRIS C. SCHÖN D., *Organizational Learning: A Theory of Action Perspective*, Addison Wesley, Reading, PA, 1978.

[ARN 98] ARNAUD P., *Chimie Physique: Cours*, Dunod, Paris, 1998.

[AUG 08] AUGENDRE M., Vivre avec le volcan: une géographie du risque volcanique au Japon, PhD Thesis, University of Lyon 2, 2008.

[AZE 07] AZEMA J., "Entretien sur les mutuelles d'assurance en France", *Revue Risques, les cahiers de l'assurance*, vol. 69, no. 1, consulté en ligne le 11 January 2016, available at: https://www.ffsa.fr/webffsa/risques.nsf/html/Risques_69_0004.htm, 2007.

[BAE 95] BAECHLER J., *Le capitalisme, Tome 1: les origines; Tome 2: l'économie capitaliste*, Gallimard, Paris, 1995.

[BAI 82] BAILLY A., BEGUIN H., *Introduction à la géographie humaine*, Masson, Paris, 1982.

[BAI 85] BAIROCH P., *De Jéricho à Mexico. Villes et économie dans l'histoire*, Gallimard, Paris, 1985.

[BAI 97] BAILLY A., FERRAS R., *Eléments d'épistémologie de la géographie*, Armand Colin, Paris, 1997 (2010).

[BAK 99] BAK P., *Quand la nature s'organise*, Flammarion, Paris, 1999.

[BAR 23] BARROWS H., "Geography as human ecology", *Annals of the Association of American Geographers*, vol. 13, no. 1, pp. 1–14, 1923.

[BAS 13] BASSETT T., FOGELMAN C., "Déjà vu or something new? The adaptation concept in the climate change literature", *Geoforum*, vol. 48, pp. 42–53, 2013.

[BEC 86] BECK U., *La société du risque. Sur la voie d'une autre modernité*, Flammarion, Paris, 1986.

[BEN 93] BENEVOLO L., *La ville dans l'histoire européenne*, Seuil, Paris, 1993.

[BER 34a] BERGSON H., *Introduction à la métaphysique*, Payot, Paris, 1934 (reprinted in 2013).

[BER 34b] BERGSON H., *La pensée et le mouvant*, Flammarion, Paris, 1934.

[BER 41] BERGSON H., *L'évolution créatrice*, Presses Universitaires de France, Paris, 1941 (reprinted in 2014).

[BER 82] BERQUE A., *Vivre l'espace au Japon*, Presses Universitaires de France, Paris, 1982.

[BER 88] BERDOULAY V., *Des mots et des lieux. La dynamique du discours géographique*, Editions du CNRS, Paris, 1988.

[BER 02] BERKES F., FOLKE C., "Back to the future: ecosystem dynamics and local knowledge", in GUNDERSON L.H., HOLLING C.S. (eds), *Panarchy: Understanding Transformations in Human and Natural Systems*, Island Press, Washington DC, 2002.

[BER 13] BERTRAND F., ROCHER L. (eds), *Les territoires face aux changements climatiques*, Peter Lang, Bern, 2013.

[BES 04] BESANCENOT J.-P., Une vague de chaleur meurtrière: les enseignements de l'été 2003 en France, available at: http://geoconfluences.ens-lyon.fr/, 2004.

[BIR 06] BIRKMANN J. (ed.), *Measuring Vulnerability to Natural Hazards. Towards Disaster Resilient Societies*, United Nations University Press, Tokyo, 2006.

[BLA 87] BLAIKIE P., BROOKFIELD H., *Land Degradation and Society*, Methuen, London and New York, 1987.

[BLA 94] BLAIKIE P., CANNON T., DAVIS I. *et al.*, *At Risk: Natural Hazards, People's Vulnerability and Disasters*, Routledge, London, 1994.

[BLA 15] BLANC P., Inventaire 2014 des accidents technologiques, available at: http://www.aria.developpement-durable.gouv.fr/wp-content/files_mf/Inventaire 2014_ web.pdf, 2015.

[BLO 98] BLONDIAUX L., *La fabrique de l'opinion: une histoire sociale des sondages*, Seuil, Paris, 1998.

[BOI 13] BOISSIER L., La mortalité liée aux crues torrentielles dans le sud de la France: une approche de la vulnérabilité humaine face à l'inondation, PhD Thesis, University of Montpellier 3, 2013.

[BOU 12] BOURFOUKA H., KRAUTBERGER N., "Objectivité, vérité, et assertabilité en histoire environnementale", *Etudes Rurales*, vol. 1, no. 189, pp. 181–198, 2012.

[BOU 15] BOUDIA S., HENRY E. (eds), *La mondialisation des risques*, Presses Universitaires de Rennes, Rennes, 2015.

[BRA 85] BRAUDEL F., *La dynamique du capitalisme*, Flammarion, Paris, 1985.

[BUR 01] BURBY R.J., NELSON A.C., PARKER D. *et al.*, "Urban containment policy and exposure to natural hazards: is there a connection?", *Journal of Environmental Planning and Management*, vol. 44, no. 4, pp. 475–490, 2001.

[CAL 12] CALVO S., Gestion du risque avalanche et pression foncière à Chamonix-Mont-Blanc. Etude de cas du Brévent, Master Thesis 1, University of Savoie, 2012.

[CAM 05] CAMBIER A., *Qu'est-ce qu'une ville?*, Vrin, Paris, 2005.

[CAR 62] CARSON R., *Silent Spring*, Houghton Mifflin Harcourt, Boston, 1962.

[CAS 14] CASTREE N., ADAMS W., BARRY J. *et al.*, "Changing the intellectual climate", *Nature Climate Change*, vol. 4, pp. 763–768, 2014.

[CHA 94] CHALINE C., DUBOIS-MAURY J., *La ville et ses dangers. Prévention et gestion des risques naturels, sociaux et technologiques*, Masson, Paris, 1994.

[CHA 14] CHANDLER D., *Resilience. The Governance of Complexity*, Routledge, London and New York, NY, 2014.

[CHE 08] CHEMLA G., BILLOT P. (eds), *L'environnement urbain à Colombo: diagnostic et enjeux*, Asia Pro Eco, Colombo, 2008.

[COA 06] COANUS T., PÉROUSE J.-F., *Villes et risques. Regards croisés sur quelques cités en danger*, Anthropos, Paris, 2006.

[COM 07] COMBE C., La ville endormie: le risque d'inondation à Lyon. Approche géohistorique et systémique du risque de crue en milieu urbain et péri-urbain, PhD Thesis, University of Lyon 2, 2007.

[COM 15] COMBY E., Pour qui l'eau? Les contrastes spatio-temporels des discours sur le Rhône (France) et le Sacramento (Etats-Unis), PhD Thesis, University of Lyon 3, 2015.

[COP 10] COPRNM, Summary of meeting on June 17th 2010, MEEDDM – DGPR, available at: http://www.developpementdurable.gouv.fr/IMG/pdf/100617_CrCoprnm.pdf, 2010.

[COR 11] CORCUFF P., "Le savant et le politique", *SociologieS*, available at: https://sociologies.revues.org/3533, 2011.

[COU 13] COUTELLEC L., *De la démocratie dans les sciences. Epistémologie, éthique et pluralisme*, Editions Matériologiques, Paris, 2013.

[CRE 15] CRED, FAQS, available at: http://www.emdat.be/frequently-asked-questions, 2015.

[CUM 11] CUMMING G., *Spatial Resilience in Social-Ecological Systems*, Springer, Dordrecht, 2011.

[CUT 96] CUTTER S.L., "Vulnerability to environmental hazards", *Progress in Human Geography*, vol. 20, no. 4, pp. 529–539, 1996.

[CUT 00] CUTTER S.L., MITCHELL J.T., SCOTT M.S., "Revealing the vulnerability of people and places: a case study of Georgetown County, South Carolina", *Annals of the Association of American Geographers*, vol. 90, no. 4, pp. 713–737, 2000.

[CUT 08] CUTTER S.L., BARNES L., BERRY M. *et al.*, "A place-based model for understanding community resilience to natural disasters", *Global Environmental Change*, vol. 18, pp. 598–609, 2008.

[CUT 13] CUTTER S.L., ZLATANOVA S., EHRLICH D., Welcome to the DATA community of practice, available at: http://www.irdrinternational.org/projects/data/, 2013.

[DAH 07] DAHAN DALMEDICO A., "Le régime climatique, entre science, expertise et politique", in DAHAN DALMEDICO A. (ed.), *Les modèles du futur. Changement climatique et scénarios économiques: enjeux scientifiques et politiques*, La Découverte, Paris, 2007.

[DAH 13] DAHAN DALMEDICO A., "Le changement climatique: l'exception d'un risqué", in BOURG D., JOLY P.-B., KAUFMANN A. (eds), *Du risque à la menace, penser la catastrophe*, Presses Universitaires de France, Paris, 2013.

[DAU 13] DAUPHINÉ A., PROVITOLO D., *Risques et catastrophes: observer, spatialiser, comprendre, gérer*, Armand Colin, Paris, 2013.

[DE 75] DE ROSNAY J., *Le macroscope. Vers une vision globale*, Seuil, Paris, 1975.

[DE 12] DE BELIZAL E., Les corridors de lahars du volcan Merapi (Java, Indonésie): des espaces entre risque et ressource, PhD Thesis, University of Paris 1, 2012.

[DE 13] DE GROEVE T., POLJANSEK K., EHRLICH D., Recording Disaster Losses: Recommendations for a European Approach, EU Commission DG JRC, available at: http://publications.jrc.ec.europa.eu/repository/bitstream/JRC83743/lbna26111enn.pdf, 2013.

[DE 14] DE GROEVE T., POLJANSEK K., EHRLICH D. *et al.*, Current Status and Best Practices for Disaster Loss Data recording in EU Member States, a comprehensive overview of current practice in the EU member states, EU Commission DG JRC, available at: http://drr.jrc.ec.europa.eu/Portals/0/Loss/JRC%20SOTA%20Loss%20Report_11182014.pdf, 2014.

[DEM 00] DEMERITT D., "The new social contract for science: accountability, relevance, and value in US and UK science and research policy", *Antipode*, vol. 32, no. 3, pp. 308–329, 2000.

[DEM 01] DEMERITT D., "The construction of global warming and the politics of science", *Annals of the Association of American Geographers*, vol. 91, no. 2, pp. 307–337, 2001.

[DEM 09] DEMERITT D., "Geography and the promise of integrative environmental research", *Geoforum*, vol. 40, no. 2, pp. 127–129, 2009.

[DER 99] D'ERCOLE R., PIGEON P., "L'expertise internationale des risques dits naturels: intérêt géographique", *Annales de Géographie*, vol. 108, no. 608, pp. 339-357, 1999.

[DER 09a] D'ERCOLE R., METZGER P., "La vulnérabilité territoriale: une nouvelle approche des risques en milieu urbain", *Cybergéo*, available at: https://cybergeo.revues.org/22022, 2009.

[DER 09b] D'ERCOLE R., HARDY S., ROBERT J., "Balance de los accidentes y desastres ocurridos en La Paz, Lima y Quito (1970–2007)", *Bulletin de l'Institut Français d'Etudes Andines*, vol. 38, no. 3, pp. 433–465, 2009.

[DES 37] DESCARTES R., *Discours de la méthode*, Union Générale d'Editions, Paris, 1637 (1951).

[DES 05] DESCOLA P., *Par-delà nature et culture*, Gallimard, Paris, 2005.

[DGP 14] DGPR, Stratégie nationale de gestion des risques d'inondation, available at: http://www.developpement-durable.gouv.fr/IMG/pdf/140509_SNGRIet Annexes_approuvee_BAT_cle0459ad.pdf, 2014.

[DOM 08] DOMENICHINI J., Calcul de la sinistralité relative à des aléas "naturels" et anthropiques rares: systèmes d'information géographique (SIG), retour d'expérience et méthodes prospectives pour l'évaluation de sinistres inondation et terrorisme en réassurance, PhD Thesis, University of Paris 1, 2008.

[DOW 09] DOWNTON P.F., *Ecopolis: Architecture and Cities for a Changing Climate*, Springer, Berlin, 2009.

[DRO 93] DROUIN J.-M., *L'écologie et son histoire*, Flammarion, Paris, 1993.

[DUN 14] DUNCAN C., SCHERER S., WADE-APICELLA S., HFA Thematic Review: Research Area 2. Priority for Action 3 – Core Indicator 1: Relevant information on disasters is available and accessible at all levels, to all stakeholders (through networks, development of information sharing systems etc.), Background Paper prepared for the 2015 Global Assessment Report on Disaster Risk Reduction, available at: http://www.preventionweb.net/english/hyogo/gar/2015/en/bgdocs/UNISDR,%202014c.pdf, UNISDR Geneva, 2014.

[DUP 05] DUPUY J.-P., *Petite métaphysique des tsunamis*, Seuil, Paris, 2005.

[DUP 08] DUPONT C., PIGEON P., Le Haut-Rhône et son bassin-versant montagneux: pour une gestion intégrée des territoires transfrontaliers, report, Institut de la montagne, Chambéry, 2008.

[EVA 14] EVANS B., REID J., *Resilient Life. The Art of Living Dangerously*, Polity Press, Cambridge, 2014.

[FAI 15] FAINSTEIN S., "Resilience and justice", *International Journal of Urban and Regional Research*, vol. 39, no. 1, pp. 157–167, 2015.

[FAT 06] FATTAL P., Sensibilité et vulnérabilité des côtes aux pollutions par les hydrocarbures, Thesis, University of Nantes, 2006.

[FEL 14] FELLI R., "Adaptation et résilience: critique de la nouvelle éthique de la politique environnementale international", *Ethique publique*, vol. 16, no. 1, available at: https://ethiquepublique.revues.org/1371, 2014.

[FEL 15] FELLI R., "Environment, not planning: the neoliberal depoliticisation of environmental policy by means of emissions trading", *Environmental Politics*, vol. 24, no. 5, pp. 641–660, 2015.

[FES 57] FESTINGER L., *A Theory of Cognitive Dissonance*, Stanford University Press, Stanford, 1957.

[FOR 03] FORSYTH T., *Critical Political Ecology. The Politics of Environmental Science*, Routledge, London – New York, 2003.

[FOR 12] FORSYTH T., "Politicizing environmental science does not mean denying climate science nor endorsing it without question", *Global Environmental Politics*, vol. 12, no. 2, pp. 18–23, 2012.

[GAI 08] GAILLARD J.-C., "Alternative paradigms of volcanic risk perception: the case of Mt. Pinatubo in the Philippines", *Journal of Volcanology and Geothermal Research*, vol. 172, nos 3–4, pp. 315–328, 2008.

[GAI 10] GAILLARD J.-C., "Vulnerability, capacity and resilience: perspectives for climate and development policy", *Journal of International Development*, vol. 22, no. 2, pp. 218–232, 2010.

[GAL 15] GALL M., NGUYEN K., CUTTER S.L., "Integrated research on disaster risk: is it really integrated?", *International Journal of Disaster Risk Reduction*, vol. 12, pp. 255–267, June 2015.

[GAR 05] GARCIA ACOSTA V., "El riesgo como construccion social y la construccion social de riesgos", *Desacatos*, no. 19, pp. 11–24, September–December 2005.

[GAR 09] GARDAIX J., Géopolitique et risques de marée noire en Europe, PhD Thesis, University of Bordeaux III, 2009.

[GAR 15] GARCIA P.-O., Sous l'adaptation, l'immunité. Etude sur le discours de l'adaptation au changement cliamtique, PhD Thesis, University of Grenoble Alpes, 2015.

[GID 98] GIDDENS A., PIERSON C., *Conversations with Anthony Giddens: Making Sense of Modernity*, Polity Press, Cambridge, 1998.

[GIL 13] GILBERT C., "Quels risques pour la recherche en sciences humaines et sociales?", in BOURG D., JOLY P.-B., KAUFMANN A. (eds), *Du risque à la menace, penser la catastrophe*, Presses Universitaires de France, Paris, 2013.

[GLA 09] GLANTZ M.H., *Heads Up! Early Warning Systems for Climate, Water and Weather Related Hazards*, United Nations University Press, Tokyo, 2009.

[GLA 15] GLANTZ M.H., The Antalya statement – An expert forum on disaster risk reduction (DRR) in a changing climate: lessons learned about lessons learned, USAID, CCB/CU, WMO, TSMS with the support of NOAA and GFDRR, Antalya, 2015.

[GOD 02] GODARD O., HENRY C., LAGADEC P. *et al.*, *Traité des nouveaux risques*, Gallimard, Paris, 2002.

[GOU 15] GOUVERNEMENT DU CANADA, Base de données canadienne sur les catastrophes, available at: https://www.publicsafety.gc.ca/cnt/rsrcs/cndn-dsstr-dtbs/index-fra.aspx, 2015.

[GRO 14] GROVE K., "Agency, affect, and the immunological politics of disaster resilience", *Environment and Planning D: Society and Space*, vol. 32, pp. 240–256, 2014.

[GTZ 05] GTZ, Policy framework and operational strategy, Participatory improvement of under-served settlements in Colombo, Final report, Colombo, 2005.

[GUE 14a] GUERRIN J., Une inondation négociée? Politisation d'un risque naturel sur le Rhône, PhD Thesis, University of Montpellier 1, 2014.

[GUE 14b] GUÉZO B., PIGEON P., "Les défis liés à la prévention des désastres dans les aires métropolitaines: exemple de Givors dans l'aire métropolitaine lyonnaise (France)", *Vertigo*, vol. 14, no. 3, available at: http://vertigo.revues.org/15842, 2014.

[GUN 02] GUNDERSON L.H., HOLLING C.S. (eds), *Panarchy: Understanding Transformations in Human and Natural Systems*, Island Press, Washington DC, 2002.

[HAL 07] HALLEGATTE S., THÉRY D., "Les impacts économiques futurs du changement climatique sont-ils sous-estimés?", *Journal of Economics Policy*, vol. 117, no. 4, pp. 507–522, 2007.

[HAR 88] HARAWAY D., "Situated knowledges: the science question in feminism and the privilege of partial perspective", *Feminist Studies*, vol. 14, no. 3, pp. 575–599, 1988.

[HAR 13] HARDY S., *Atlas de la vulnérabilité de l'agglomération de La Paz*, IRD, Paris, 2013.

[HOL 73] HOLLING C.S., "Resilience and stability of ecological systems", *Annual Review of Ecology and Systematics*, vol. 4, pp. 1–23, 1973.

[HUL 06] HULBERT F. (ed.), *Villes du nord, villes du sud: géopolitique urbaine, acteurs et enjeux*, L'Harmattan, Paris, 2006.

[HUL 09] HULME M., "Why we disagree about climate change", *Understanding Controversy, Inaction and Opportunity*, Cambridge University Press, Cambridge, MA, 2009.

[IKM 13] IKM4DRR, Draft – Workshop, global platform for disaster risk reduction, 20 May 2013, unpublished, available at: http://www.preventionweb.net/files/workspace/33381_ikm4drrframeworkdraft200513.pdf, 2013.

[IPC 12] IPCC, "Managing the risks of extreme events and disasters to advance climate change adaptation", *Summary for Policymakers*, Cambridge University Press, Cambridge, MA, 2012.

[IRD 15] IRDR (Integrated Research on Disaster Risk), Annual report 2014, addressing the challenges of natural & human-induced environmental hazards, available at: http://www.wcdrr.org/wcdrr-data/uploads/579/IRDR%20Annual%20Report%202014.pdf, 2015.

[ISE 14] ISEMAR, La sécurité dans la logistique maritime et portuaire des vracs, summary note no. 161, available at: http://www.isemar.asso.fr/fr/pdf/note-de-synthese-isemar-161.pdf, January 2014.

[ITO 15] ITOPF, Oil tanker spill statistics 2015, available at: http://www.itopf.com/knowledge-resources/data-statistics/statistics/, 2015.

[JAN 06] JANSSEN M.A., OSTROM E., "Resilience, vulnerability, and adaptation: a cross-cutting theme of the International Human Dimensions Programme on Global Environmental Change", *Global Environmental Change*, vol. 16, pp. 237–239, 2006.

[JAN 14] JANKOVIC V., BOWMAN A., "After the green gold rush: the construction of climate change as a market transition", *Economy and Society*, vol. 43, no. 2, pp. 233–259, 2014.

[JON 90] JONAS H., *Le principe responsabilité*, Flammarion, Paris, 1990.

[JOU 06] JOUVE B., "L'empowerment: entre mythe et réalités, entre espoirs et désenchangement", *Géographie, économie, société*, vol. 8, pp. 5–15, 2006.

[KAL 05] KALINOWSKI I., *Leçons wébériennes sur la science et la propagande*, Agone, Marseille, 2005.

[KAT 86] KATES R.W., BURTON I., "Geography, resources and environment", *Selected Writings of Gilbert F. White*, University of Chicago Press, Chicago, IL, 1986.

[KEL 08] KELMAN I., GAILLARD J.C., "Placing climate change within disaster risk reduction", *Disaster Advances*, vol. 1, no. 3, pp. 3–5, 2008.

[KEL 15a] KELMAN I., GLANTZ M., "Analysing the Sendai framework for disaster risk reduction", *International Journal for Disaster Risk Science*, vol. 6, pp. 105–106, 2015.

[KEL 15b] KELMAN I., GAILLARD J.C., MERCER J., "Climate change's role in disaster risk reduction's future: beyond vulnerability and resilience", *International Journal of Disaster Risk Science*, vol. 6, no. 1, pp. 21–27, 2015.

[KUH 10] KUHLICKE C., STEINFUHRER A., Social capacity building for natural hazards: a conceptual frame, Cap-Haz net report, Helmholtz Centre for Environmental Research – UFZ, Leipzig, 2010.

[LA 02] LA RED, Corporación OSSO, UNISDR, Comparative analysis of disaster databases, available at: http://www.desinventar.org/en/proyectos/lared/comparacion/index.html, 2002.

[LAT 94] LATOUR B., *Nous n'avons jamais été modernes. Essai d'anthropologie symétrique*, La Découverte, Paris, 1994.

[LAV 13] LAVELL A., MASKREY A., The future of disaster risk management: an ongoing discussion, UNISDR-FLACSO, available at: https://www.unisdr.org/we/inform/publications/35715, 2013.

[LE 13] LE TREUT H. (ed.), *Les impacts du changement climatique en Aquitaine*, Aquitaine, Bordeaux, 2013.

[LÓP 08] LÓPEZ J., La construction sociale du risque à Medellín (Colombie): Gouvernance locale et représentations, PhD Thesis, School of Advanced Social Science Studies (EHESS), Paris, 2008.

[LÓP 11] LÓPEZ-PELÁEZ J., PIGEON P., "Co-evolution between structural mitigation measures and urbanization in France and Colombia: a comparative analysis of disaster risk management policies based on disaster databases", *Habitat International*, vol. 35, pp. 573–581, 2011.

[LÖV 15] LÖVBRAND E., BECK S., CHILVERS J. *et al.*, "Who speaks for the future of Earth? How critical social science can extend the conversation on the Anthropocene", *Global Environmental Change*, vol. 32, pp. 211–218, 2015.

[MAC 12] MACCAFERRI S., CARIBONI F., CAMPOLONGO F., Natural catastrophes: risk relevance and insurance coverage in the EU, European Commission Joint Research Centre, Ispra (Italy), available at: http://ec.europa.eu/internal_market/insurance/docs/natural-catastrophes/jrc_report_on_nat_cat_en.pdf, 2012.

[MAD 13] MADAVAN D., Les minorités tamoules à Colombo, Kuala Lumpur et Singapour: minorités, intégrations socio-spatiales et citadinités, PhD Thesis, University of Paris IV Sorbonne, 2013.

[MAN 06] MANCEBO F., "Du risque 'naturel' à la catastrophe urbaine: Katrina", *Vertigo*, vol. 7, no. 1, available at: http://vertigo.revues.org/2096, 2006.

[MAS 93] MASKREY A. (ed.), Los desastres no son naturales, *La Red*, available at: http://www.oei.es/decada/portadas/Desnat.pdf, 1993.

[MC 07] MC ENTIRE D.A., *Disciplines, Disasters and Emergency Management: The Convergence and Divergence of Concepts, Issues and Trends from the Research Literature*, Charles C Thomas Publisher, Springfield, 2007.

[MEA 72] MEADOWS D., MEADOWS D., RANDERS J. *et al.*, *The Limits to Growth*, Universe Books, New York, 1972.

[MEN 11] MENONI S., MARGOTTINI C. (eds), *Inside Risk: A Strategy for Sustainable Risk Mitigation*, Springer, Berlin, 2011.

[MER 10] MERCER J., "Disaster risk reduction or climate change adaptation: are we reinventing the wheel?", *Journal of International Development*, vol. 22, pp. 247–264, 2010.

[MET 11] METZGER P., D'ERCOLE R., "Les risques en milieu urbain: éléments de réflexion", *EchoGéo*, no. 18, available at: http://echogeo.revues.org/12640, 2011.

[MET 15] METZGER P., ROBERT J., "Environnement et risques: les sciences sociales piégées entre critique radicale et utilité sociale", in CLERVAL A., LEBLANC A., REBOTIER J. *et al.* (eds), *Espace et rapports de domination*, Presses Universitaires de Rennes, Rennes, 2015.

[MIC 13] MICHEL-KERJEAN E., "Finance des risques catastrophiques: le marché américain est en plein bouleversement", *Revue économique*, vol. 64, no. 4, pp. 615–634, 2013.

[MIT 14] MITCHELL T., GUHA-SAPIR D., HALL J. *et al.*, Setting, measuring and monitoring targets for reducing disaster risk. Recommendations for post-2015 international policy frameworks, Overseas Development Institute, London, available at: http://www.odi.org/publications/8448-setting-measuring-monitoring-targets-disaster-risk-reduction-recommendations-post-2015-international-policy-frameworks, 2014.

[MOA 15] MOATTY A., Pour une géographie des reconstructions post-catastrophe: risques, sociétés, territoires, PhD Thesis, University of Montpellier 3, 2015.

[MOC 14] MOCHIZUKI J., "Decision-making, policy choices and community rebuilding after the Tohoku disaster", *IDRiM Journal*, vol. 4, no. 2, available at: http://www.idrimjournal.com/index.php/idrim/article/view/85, 2014.

[MOR 82] MORIN E., *Science avec conscience*, Fayard, Paris, 1982.

[MOR 86] MORIN E. *La Méthode. 3, la connaissance de la connaissance*, Seuil, Paris, 1986.

[MOR 05] MORIN E. *Introduction à la pensée complexe*, Seuil, Paris, 2005.

[MOR 11] MORIN E., *Mes philosophes*, Germina, Paris, 2011.

[MOU 14] MOUGIN P., Les torrents de la Savoie, Société d'histoire naturelle de Savoie, Grenoble, 1914.

[MOU 08] MOULIN A., PIGEON P., Inondation et urbanisation à Colombo (Sri Lanka): apports de l'analyse systémique en géographie, conférence finale du programme COLEM, Colombo, 2008.

[MUD 00] MUDHC, Action plan for the control of wastewater disposal into the greater Colombo canal system, Greater Colombo Flood Control and Environment Project, Phase 2, Nippon Koei and Atkins international, Colombo, 2000.

[MUS 02] MUSSET A., *Villes nomades du Nouveau Monde*, Editions de l'EHESS, Paris, 2002.

[NAT 15] NATIONS UNIES, Cadre de Sendai pour la réduction des risques de catastrophe (2015–2030), Nations-Unies, New York, available at: http://www.unisdr.org/files/43291_frenchsendaiframeworkfordisasterris.pdf, 2015.

[NOB 14] NOBERT S., REBOTIER J., Résilience: une histoire temporelle oubliée au cœur de la gouvernance des futurs, Délivrable Conceptuel Programme ACTER, unpublished, 2014.

[NUS 15] NUSSBAUM R., PIGEON P., "A national public private partnership (PPP) platform for risk data sharing to stimulate participative governance in France", UNISDR Scientific and Technical Advisory Group, Case Studies, available at: http://www.preventionweb.net/files/workspace/7935_rnussbaumpppdrrinfrance.pdf, 2015.

[OBR 07] O'BRIEN K.L., ERIKSEN S., NYGAARD L. *et al.*, "Why different interpretations of vulnerability matter in climate change discourses", *Climate Policy*, vol. 7, pp. 73–88, 2007.

[OCH 15] OCHOA E.,"Algunas reflexiones sobre las dimensiones político institucionales de la gestión de riesgos en Ecuador", in BERMÚDEZ D.N., REBOTIER J. (eds), *Dimensiones político institucionales de la gestión de riesgos y vulnerabilidades en los países andinos. Reflexiones y avances*, PNUD, Quito, 2015.

[OKE 76] O'KEEFE P., WESTGATE K., WISNER B., "Taking the naturalness out of natural disasters", *Nature*, vol. 260, no. 5552, pp. 566–567, 1976.

[OLC 16] OLCINA J., SAURI D., HERNÁNDEZ M. *et al.*, "Flood policy in Spain: a review for the period 1983–2013", *Disaster Prevention and Management*, vol. 25, no. 1, pp. 41–58, 2016.

[PAH 09] PAHL-WOSTL C., "A conceptual framework for analysing adaptive capacity and multi-level learning processes in resource governance regimes", *Global Environmental Change*, vol. 19, pp. 354–365, 2009.

[PAL 13] PALLETT H., CHILVERS J., "A decade of learning about publics, participation and climate change", *Environment and Planning A*, vol. 45, no. 5, pp. 1162–1183, 2013.

[PAS 63] PASCAL B., *Oeuvres complètes*, Seuil, Paris, 1963.

[PAS 95] PASSERON J.-C., "L'espace mental de l'enquête (I). La transformation de l'information sur le monde dans les sciences sociales", *Enquête*, vol. 1, pp. 13–42, available at: http://enquete.revues.org/259, 1995.

[PAS 96] PASSERON J.-C., "La constitution des sciences sociales", *Le Débat*, vol. 3, no. 90, pp. 93–112, 1996.

[PEY 95] PEYREFITTE A., *La société de confiance. Essai sur les origines et la nature du développement*, Odile Jacob, Paris, 1995.

[PIC 13] PICQOUT A., Impacts géographiques de l'éruption de 2010 du volcan Merapi, Java, Indonésie, PhD Thesis, University of Paris 1, 2013.

[PIG 91] PIGEON P., L'homme face au risque lié aux glissements de terrain dans le massif du Chablais et son piedmont, PhD Thesis, University of Grenoble 1, 1991.

[PIG 94] PIGEON P., *Ville et environnement*, Nathan, Paris, 1994.

[PIG 02] PIGEON P., Réflexions sur la géographie des risques dits naturels, Mémoire d'Habilitation à Diriger des Recherches, University of Nice-Sophia-Antipolis, 2002.

[PIG 05] PIGEON P., *Géographie critique des risques*, Economica, Paris, 2005.

[PIG 07] PIGEON P., *L'environnement au défi de l'urbanisation*, Presses universitaires de Rennes, Rennes, 2007.

[PIG 08] PIGEON P., "L'évaluation des risques à Colombo: résultats d'une approche systémique", in CHEMLA D.G., BILLOT P. (eds), *L'environnement urbain à Colombo, diagnostic et enjeux*, Asia Pro Eco, Colombo, 2008.

[PIG 10] PIGEON P., "Catastrophes dites naturelles, risques et développement durable: utilisations géographiques de la courbe de Farmer", *Vertigo*, vol. 10, no. 1, available at: https://vertigo.revues.org/9491, 2010.

[PIG 12a] PIGEON P., *Paradoxes de l'urbanisation. Pourquoi les catastrophes n'empêchent-elles pas l'urbanisation?* L'Harmattan, Paris, 2012.

[PIG 12b] PIGEON P., "Apports de la résilience à la géographie des risques: l'exemple de La Faute-sur-Mer (Vendée, France)", *Vertigo*, vol. 12, no. 1, available at: https://vertigo.revues.org/12031, 2012.

[PIG 14] PIGEON P., "Risque digue: une justification à la relecture systémique et géopolitique des risques environnementaux", *L'Espace Politique*, vol. 24, no. 3, available at: https://espacepolitique.revues.org/3256, 2014.

[PIG 15] PIGEON P., Keynote document highlighting some barriers to implementation of already mature DRR policies in the EU, KNOW 4 DRR Deliverable, 2015.

[PIG 16] PIGEON P., "Dike risk: revealing the academic links between disaster risk reduction, sustainable development, climate change and migration", SUDMEIER-RIEUX D.K., FERNÁNDEZ M., PENNA I. *et al.* (eds), *Identifying Emerging Issues in Disaster Risk Reduction, Migration, Climate Change and Sustainable Development. Shaping Debates and Policies*, Springer, London, 2016.

[POP 73] POPPER K., *La Logique de la découverte scientifique*, Payot, Paris, 1973.

[POR 09] PORTAL T. (ed.), *Crises et facteur humain. Les nouvelles frontières mentales des crises*, De Boeck, Brussels, 2009.

[PRE 14] Préfecture de la Gironde, Révision du Plan de Prévention des Risques d'Inondation (PPRI) de la Garonne sur les secteurs de Saint Pierre d'Aurillac à La Réole, available at: http://www.gironde.gouv.fr/Politiques-publiques/Environnement-risques-naturels-et-technologiques/Prevention-des-risques/L-Etat-face-aux-risques/, 2014.

[PRO 15a] PROPECK-ZIMMERMANN E., Modélisation des risques et décision territoriale: recherches sur les risques socio-techniques en milieu urbain, University of Caen, 2015.

[PRO 15b] PROVITOLO D., REGHEZZA-ZITT M., "Resilience and vulnerability: from opposition towards a continuum", in REGHEZZA-ZITT D.M., RUFAT S. (eds), *Resilience Imperative*, ISTE Press Ltd. London and Elsevier Ltd. Oxford, 2015.

[QUA 06] QUARANTELLI E., Catastrophes are different from disasters: some implications for crisis planning and managing drawn from Katrina, Understanding Katrina, Perspectives from social sciences, Social Science Research Council, available at: http://understandingkatrina.ssrc.org/Quarantelli/, 2006.

[QUE 13] QUENAULT B., "Retour critique sur la mobilisation du concept de résilience en lien avec l'adaptation des systèmes urbains au changement climatique", *EchoGéo*, vol. 24, available at: https://echogeo.revues.org/13403, 2013.

[QUE 15] QUENAULT B., "De Hyōgo à Sendai, la résilience comme impératif d'adaptation aux risques de catastrophe: nouvelle valeur universelle ou gouvernement par la catastrophe?", *Développement durable et territoires*, vol. 6, no. 3, available at: https://developpementdurable.revues.org/11010, 2015.

[RAM 10] RAMÍREZ GALLEGOS F., "Post-neoliberalismo indócil. Agenda pública y relaciones socio-estatales en el Ecuador de la Revolución Ciudadana", *Temas y Debates*, vol. 14, no. 20, pp. 175–194, 2010.

[REB 12] REBOTIER J., "Vulnerability conditions and risk representations in Latin-America: framing the territorializing urban risk", *Global Environmental Change*, vol. 22, no. 2, pp. 391–398, 2012.

[REB 13a] REBOTIER J., "Enjeux et défis des politiques locales d'adaptation au changement climatique en Aquitaine", *Geographicalia*, vol. 63–64, pp. 157–176, 2013.

[REB 13b] REBOTIER J., LOPEZ-PELAEZ J., PIGEON P., "Las paradojas de la resiliencia: miradas cruzadas entre Colombia y Francia", *Territorios*, vol. 28, pp. 127–145, 2013.

[REB 14] REBOTIER J., "Adaptations aux changements environnementaux et territoires. Questions de (science en) société", *Sud-Ouest Européen*, vol. 37, pp. 5–12, 2014.

[REB 15] REBOTIER J., "Politiques de gestion des risques en Equateur. Entre coopération internationale et réaffirmation de l'Etat", *Autrepart*, vol. 74, 2015.

[REB 16] REBOTIER J., *El riesgo y su gestión en Ecuador. Una mirada de geografía social y política*, Ediciones de la PUCE–IRD, Quito, 2016.

[REG 15a] REGHEZZA-ZITT M. "Territorialiser ou ne pas territorialiser le risque et l'incertitude. La gestion territorialisée à l'épreuve du risque d'inondation en Île-de-France", *L'Espace Politique*, vol. 26, no. 2, available at: https://espacepolitique.revues.org/3543, 2015.

[REG 15b] REGHEZZA-ZITT M., RUFAT S. (eds), *Resilience Imperative*, ISTE Press Ltd, London, Elsevier Ltd, Oxford, 2015, 2015.

[REG 15c] REGHEZZA-ZITT M., RUFAT S., "L'adaptation en Île-de-France entre injonction et recyclage. Techniques et politiques de la société de l'incertitude", *Développement durable et territoires*, vol. 6, no. 3, available at: https://developpementdurable.revues.org/11035, December 2015.

[REN 13] RENAUD F., SUDMEIER-RIEUX K., ESTRELLA M. (eds), *The Role of Ecosystems in Disaster Risk Reduction*, United Nations University Press, Tokyo, 2013.

[REV 02] REVAULT D'ALLONNES M., *Le dépérissement de la politique. Généalogie d'un lieu commun*, Flammarion, Paris, 2002.

[REV 11] REVET S., "Penser et affronter les désastres: un panorama des recherches en sciences sociales et des politiques internationales", *Critique Internationale*, vol. 52, no. 3, pp. 157–173, 2011.

[RIB 10] RIBOT J., "Vulnerability does not fall from the sky. Towards a multiscale, pro-poor climate policy", in MEARNS R., NORTON A. (eds), *Social Dimensions of Climate Change*, The World Bank, Washington DC, pp. 47–74, 2010.

[RIV 14] RIVERA C., "Integrating climate change adaptation into disaster risk reduction in urban contexts: perceptions and practice", *PLOS Currents Disasters*, available at: http://currents.plos.org/disasters/, January 2014.

[ROB 04] ROBBINS P., *Political Ecology. Critical Introductions to Geography*, Blackwell Publishing, Malden, 2004.

[ROB 09] ROBERT J., D'ERCOLE R., PIGEON P. *et al.*, "Complejidad, incertidumbre y vulnerabilidad: el riesgo asociado al volcán Cotopaxi (Quito, Ecuador)", *Bulletin de l'Institut Français d'Études Andines*, vol. 38, no. 3, pp. 709–733, available at: http://www.ifeanet.org/publicaciones/boletines/38%283%29/709.pdf, 2009.

[ROC 14] ROCKSTROM J., FALKENMARK M., FOLKE K. *et al.*, *Water Resilience for Human Prosperity*, Cambridge University Press, Cambridge, 2014.

[RUF 11] RUFAT S., Critique de la résilience pure, available at: http://hal.archives-ouvertes.fr/hal-00693162, 2011.

[RUM 10] RUMPALA Y., *Développement durable. Ou le gouvernement du changement total*, Le bord de l'eau, Lormont, 2010.

[SCH 42] SCHUMPETER J., *Capitalisme, socialisme et démocratie*, Payot, Paris, available at: http://classiques.uqac.ca/classiques/Schumpeter_joseph/ capitalisme_socialisme_demo/capitalisme.html, 1942.

[SEN 09] SEN A., *L'idée de justice*, Flammarion, Paris, 2009.

[SEN 13] SENPLADES, Situación actual de la planificación de desarrollo y ordenamiento territorial, Internal memorandum, Quito, 2013.

[SHI 13] SHI P., JAEGER G., YE Q., *Integrated Risk Governance. Science Plan and Case Studies of Large-scale Disasters*, Springer, London, 2013.

[SHI 14] SHI P., KASPERSON R. (eds), *World Atlas of Natural Disaster Risk*, Springer, New York, 2014.

[SHO 10] SHOVE E., "Beyond the ABC: climate change policy and theories of social change", *Environment and Planning A*, vol. 42, pp. 1273–1285, 2010.

[SIM 14] SIMONET G., "L'adaptation, champ de recherche à part entière?", *Sud-Ouest Européen*, no. 37, pp. 13–14, 2014.

[SIM 15] SIMONET G., FATORIC S., "Does 'adaptation to climate change' mean resignation or opportunity?", *Regional Environmental Change*, vol. 16, no. 3, pp. 789–799, April 2015.

[SLL 93] SLLRDC, "Surface rain water drainage and water stagnation in the city of Colombo", *Status of Flooding in CMC Area*, Colombo, 1993.

[SOR 11] SORENSEN A., OKATA J., *Megacities*, Springer, Tokyo, 2011.

[SPI 15] SPIEKERMANN R., KIENBERGER S., NORTON J. *et al.*, "The disaster-knowledge matrix – reframing and evaluating the knowledge challenges in disaster risk reduction", *International Journal of Disaster Risk Reduction*, vol. 13, pp. 96–108, 2015.

[SUD 14] SUDMEIER-RIEUX K., "Resilience – an emerging paradigm of danger or of hope?", *Disaster Prevention and Management*, vol. 23, no. 1, pp. 67–80, 2014.

[SUM 10] SUMI A., FUKUSHI K., HIRAÚATSU A. (eds), *Adaptation and Mitigation Strategies for Climate Change*, Springer, Tokyo, 2010.

[SUP 10] SUPIOT A., *L'esprit de Philadelphie, la justice sociale face au marché total*, Seuil, Paris, 2010.

[SUP 13] SUPPASRI A., MAS E., IMAMURA F., Field guide of tsunami damage and reconstruction site visit in Miyagi prefecture, Onagawa, Ishinomaki, Sendai, Natori, Iwanuma, available at: http://www.tsunami.civil.tohoku.ac.jp/hokusai3/J/people/member/anawat/Field%20guide/Onagawa_Iwanuma.pdf, 2013.

[SWI 03] SWISS RE, Swiss Re *sigma* preliminary estimates of catastrophe losses: despite natural catastrophes, 2003 will go down as a moderate year for property insurers, Zurich, available at: http://media.swissre.com/documents/pr_20031216_ sigma_en.pdf, 2003.

[THO 93] THOM R., *Prédire n'est pas expliquer*, Flammarion, Paris, 1993.

[TOV 69] TOVALOP, "Tanker owners voluntary agreement concerning liability for oil pollution", *International Legal Materials*, vol. 8, no. 3, pp. 497–501, 1969.

[TRE 13] TREFON T., Precariousness and well-being in Goma: a personal and narrative approach. Communication présentée lors du colloque de Gisenyi, dans le cadre du programme AVCOR, session 6, risk, vulnerability and society issues, unpublished, available at: http://avcor2013.africamuseum.be/, 2013.

[TUR 03] TURNER B., KASPERSON R., MATSON P. *et al.*, "A framework for vulnerability analysis in sustainability science", *Proceedings of the National Academy of Sciences*, vol. 100, no. 14, pp. 8074–8079, 2003.

[UNC 14] UNCTAD Handbook of Statistics 2014, available at: http://unctad.org/en/PublicationsLibrary/tdstat39_en.pdf

[UNC 15] UNCTAD Handbook of Statistics 2015, 398 Pages, available at: http://unctad.org/en/pages/PublicationWebflyer.aspx?publicationid=1414

[UNI 09] UNISDR, Terminologie pour la prévention des risques de catastrophe, Genève, UNISDR, available at: http://www.unisdr.org/files/7817_UNISDR TerminologyFrench.pdf, 2009.

[UNI 15a] UNISDR, Sendai framework for disaster risk reduction 2015–2030, UNISDR, Geneva, available at: http://www.unisdr.org/we/coordinate/sendai-framework, 2015.

[UNI 15b] UNISDR, Global assessment report on disaster risk reduction. Making development sustainable: the future of disaster risk management, UNISDR, Genève, available at: https://www.unisdr.org/we/inform/publications/42809, 2015.

[VEY 05] VEYRET Y., REGHEZZA M., "Aléas et risques dans l'analyse géographique", *Annales des Mines*, vol. 40, pp. 61–70, 2005.

[VID 04] VIDAL-ROSSET J., *Qu'est-ce qu'un paradoxe?*, Vrin, Paris, 2004.

[WAL 77] WALLISER B., *Systèmes et modèles: introduction critique à l'analyse des systèmes*, Seuil, Paris, 1977.

[WAL 06] WALKER B., SALT D., *Resilience Thinking. Sustaining Ecosystems and People in a Changing World*, Island Press, Washington DC, 2006.

[WAL 11] WALKER J., COOPER M., "Genealogies of resilience from systems ecology to the political economy of crisis adaptation", *Security Dialogue*, vol. 42, no. 2, pp. 143–160, 2011.

[WAR 13] WARNER J.F., VAN BUUREN A., EDELENBOS J. (eds), *Making Space for the River: Governance Experiences with Multifunctional River Flood Management in the US and Europe*, IWA Publishing, London, New York, NY, 2013.

[WEB 63] WEBER M., *Le savant et le politique*, Plon, Paris, 1963.

[WEI 14] WEICHSELGARTNER J., KELMAN I., "Geographies of resilience: challenges and opportunities of a descriptive concept", *Progress in Human Geography*, vol. 39, no. 3, pp. 249–267, 2014.

[WEI 15] WEICHSELGARTNER J., PIGEON P., "The role of knowledge in disaster risk reduction", *International Journal of Disaster Risk Sciences*, vol. 6, pp. 107–116, 2015.

[WHI 45] WHITE G.F., *Human Adjustment to Floods. A Geographical Approach to the Flood Problem in the United States*, University of Chicago, 1945.

[WHI 01] WHITE G.F., KATES R.W., BURTON I., "Knowing better and losing even more: the use of knowledge in hazard management", *Global Environmental Change Part B: Environmental Hazards*, vol. 3, nos 3–4, pp. 81–92,2001.

[WIS 04] WISNER B., BLAIKIE P., CANNON T. *et al.* (eds), *At Risk, Natural Hazards, People's Vulnerability and Disasters*, Routledge, London, 2004.

[WOR 93] WORLD BANK, Greater Colombo wastewater and sanitation master plan, Final report, CMC, Colombo, 1993.

[YOU 06] YOUNG O.R., BERKHOUT F., GALLOPIN G.C. *et al.*, "The globalization of socio-ecological systems: an agenda for scientific research", *Global Environmental Change*, vol. 16, no. 3, pp. 304–316, 2006.

[ZAJ 01] ZAJDENWEBER D., "Valeurs économiques et sinistres extrêmes d'origine humaine", *Revue Risques, les cahiers de l'assurance*, vol. 48, no. 3, available at: https://www.ffsa.fr/webffsa/risques.nsf/html/Risques_48_0019.htm, 2001.

[ZIA 15] ZIA A., WAGNER C.H., "Mainstreaming early warning systems in development and planning processes: multilevel implementation of Sendai framework in Indus and Sahel", *International Journal of Disaster Risk Sciences*, vol. 6, pp. 189–199, 2015.

Index

A, B, C

D, E, F

H, I, K

M, P, R

S, U, V

CPI Antony Rowe
Eastbourne, UK
June 25, 2019